RSPB GUIDE TO
BIRDSONG

Adrian Thomas

giving
nature
a home

RSPB GUIDE TO
BIRDSONG

Adrian Thomas

BLOOMSBURY WILDLIFE
LONDON · OXFORD · NEW YORK · NEW DELHI · SYDNEY

BLOOMSBURY WILDLIFE
Bloomsbury Publishing Plc
50 Bedford Square, London, WC1B 3DP, UK

BLOOMSBURY, BLOOMSBURY WILDLIFE and the Diana logo are trademarks of
Bloomsbury Publishing Plc

First published in the United Kingdom, 2019

A catalogue record for this book is available from the British Library

If you are unable to play CDs or have any issues saving the audio from the CD to your computer, you can also download the
audio from the CD here: www.bloomsbury.com/rspb-guide-to-birdsong

Library of Congress Cataloguing-in-Publication data has been applied for

ISBN: PB: 978-1-4729-5587-6; ePDF: 978-1-4729-5589-0; Enhanced ePub: 978-1-4729-5588-3

2 4 6 8 10 9 7 5 3 1

Design by Julie Dando, Fluke Art
Printed and bound in China by RR Donnelley

MIX
Paper from
responsible sources
FSC® C144853

To find out more about our authors and books visit www.bloomsbury.com and sign up for our newsletters

giving
nature
a home

For all items sold, Bloomsbury Publishing will donate a minimum of 2% of the publisher's receipts from sales of
licensed titles to RSPB Sales Ltd, the trading subsidiary of the RSPB. Subsequent sellers of this book are
not commercial participators for the purpose of Part II of the Charities Act 1992.

Contents

PART 1
An Introduction to Bird Songs and Calls

Introduction

When I was a child, my dad was the volunteer warden of a woodland nature reserve in Worcestershire. Each year in early May, he would lead guided walks in the evening. As each gathering of strangers made their way along the darkening paths, listening to the Robins and Song Thrushes sing their final refrains of the day, we all shared a palpable sense of anticipation.

When we reached our destination – a secluded glade – we would stand as though on hallowed ground, waiting, holding our breaths. At last, the main act would begin, cutting through the darkness: the song of the Nightingales (above). If luck were with us, they would perform with such power, control and creativity that people would … well, I don't use this word often: they'd swoon.

Such experiences remind me how lucky we are. It seems little short of a miracle that one group of wildlife – the birds – have evolved to make such elaborate, musical sounds that, to our ears, are so fascinating, so entrancing and at times so exquisitely beautiful. Frogs croak, bees hum and grasshoppers chirrup but none of them pour forth their soul in such an ecstasy (as Keats extolled in his 'Ode to a Nightingale'). Birdsong is the natural soundtrack to our lives; it can evoke a sense of place, a moment, a time of day, a year or a season.

However, actually getting to grips with birdsong and recognising which bird is making which sound can be seen as something of a dark art: impenetrable, unfathomable, a foreign language beyond our comprehension. Unable to make sense of it, too many people let birdsong wash over them.

But having spent more than 40 years of my life engrossed in the world of birds, I felt that there was more that could be done to help people tune in to their sounds and decode their secrets. The rewards for the effort are enormous. Those who take the time to learn will notice *many* more birds. By hearing more, you see more and understand more. Learning these sounds can give you insight into how each bird is feeling, and what it is trying to tell other birds around it. Bird sounds enliven your day and are a pleasant distraction from the daily grind.

More than anything, as with those evening Nightingale walks, birdsong can bring you joy, and that is what I really wanted to share. So I got out my old tent from the loft, picked up my rather large microphone, headed out into beautiful, wild places across the UK, and this book and the accompanying recordings are the results.

How to use this guide

Over the years I've met a few people who seem to possess a supernatural ability to hear the faintest squeak and whistle from birds that would otherwise go unseen and unnoticed, *and* to know what bird was making the sound in an instant.

On the other hand, I've met many, many more for whom bird sounds are a real challenge. Indeed, some people just don't know where to start. This book and its accompanying recordings are for all of you! See page 250 for information on how to access your complimentary digital download of the recordings.

The book is divided into three parts. The first two parts are akin to a learning course, to be taken at your own speed, and they run in tandem with the recordings and narration.

Part 1: An Introduction to Bird Songs and Calls is all about training your ear in how to listen, and giving you the vocabulary and techniques to describe any bird sound. At this stage, don't worry which bird is making which sound – that will come later.

In **Part 2: A Guide to Common Bird Songs and Calls**, we identify 65 species of bird based on their songs and calls. This will help you to recognise the vast majority of sounds that we hear in towns and villages, woodland and the countryside.

So find 5–10 minutes every day or so, sit down somewhere quiet, either with the book or the recordings or preferably both together. Then work your way through, track by track, reading the text in these first two parts of the book, listening to the narration, replaying each recording as often as you need to and moving on when you feel ready.

Finally, **Part 3** is a detailed **Reference Guide**, covering more than 250 widespread breeding, wintering and migrant bird species. You can dip into this section as and when you need to. As an additional tip, the sounds for these species and almost any other bird in the world can be found at your leisure on the excellent xeno-canto.org website.

Let's start at the very beginning…

Tracks 1 & 2

It might be tempting to skip the introductory parts of this book and get straight to the species-by-species sections. However, I encourage you to start with the opening chapters and the corresponding tracks on the recording. Get these basics in place, and everything else will be that much easier.

As the very first thing of all, listen to Tracks 1 and 2, the latter being precisely 1 minute of bird songs and calls. Play them a few times. Make a mental note of how many species you recognise; jot them down if you like. This is your starting point – you will return to this track at the end of the recordings, and hopefully, then you will discover how much you have progressed.

What is sound?

We start by going right back to basics. Don't worry; it's going to be brief and not too technical. Just consider for a couple of moments the question 'What is sound?' What is that invisible thing that comes sailing through the air, travels into your ears and which your brain translates as words, music, roars, rustles…?

A sound occurs when you twang a guitar string, start up a car engine, knock two stones together or scrunch up a sheet of paper. In every such instance, the action causes vibrations in the air. That's all that a sound is – air particles that are knocked into each other in a chain reaction, rippling out in all directions. We call those ripples sound waves.

If we are within earshot – in other words, if the sound waves reach us before they run out of energy – they are funnelled into our inner ears by the useful flappy bits of our ears that we have on the outside of our heads.

Once inside the ear, the waves hit a stretched membrane, the eardrum, causing it to vibrate which sets off a chain reaction that wobbles the hammer bone, which knocks into the anvil bone, which shakes the stirrup bone.

by the vibrations, triggering electrical impulses that dash off to tell the brain. The bigger the vibrations, the louder the sound; and the quicker vibrations, the higher the sound. It is then the job of different parts of the brain to register and interpret these signals and try and make sense of them or ignore them.

At this point, I need to add one crucial piece of information. We can only hear sounds where the air is vibrating at least 20 times a second. For example, a bumblebee wing vibrates about 180 times a second, which is why you can hear a buzzing sound. Waving your hand around also produces sound waves but it doesn't make the air vibrate quickly enough for us to detect a sound from it.

What's incredible is how sensitive our ears are. We can literally hear a pin drop. We can also detect very small differences in incoming sound waves. Imagine how many different people you can identify by their voice alone, even if they say exactly the same sentence. Or think of the number of pieces of music you could name after hearing only the opening few notes or chords.

Having a basic understanding of sound and hearing is useful in many ways when it comes to learning bird sounds, but for now, the main thing to remember is that nature has blessed most people with brilliant hearing apparatus and ability. When it comes to learning bird sounds, you can feel confident that you've got what it takes to master it.

Hammer bone | Anvil bone | Stirrup bone | Ear canal | Eardrum | Cochlea

Finally, the vibrations enter a spiral tube called the cochlea. On its inner walls are thousands of microscopic hair cells that move in the breeze made

How do birds make noises?

You only have to wander through the countryside in spring to realise that birds make incredibly varied and complex, and often loud, noises. So what 'wonder instrument' do they use?

To answer that question, it helps to consider how we do it; after all, we're pretty good at making vocal sounds ourselves. Our technique is to squeeze air from our lungs out through a narrow gap between two flaps of tissue at the top of our windpipe – our voicebox, or larynx. The flap margins vibrate really quickly, between 85 and 1,000 times a second. We adeptly change the shape of the gap to alter the sound that emerges.

Secrets of the syrinx

It had long been known that birds make their noises further down the windpipe than we do, at the point where it splits to go into the lungs. This upside-down Y-shaped section of tube is called the syrinx, the Greek word meaning 'pan pipes'.

It took until the twenty-first century, however, to discover the actual mechanism: bird sounds are created as air passes between pairs of flaps that project from the syrinx walls. Sound familiar? Yes, birds and humans create sound vibrations in a surprisingly similar way, just in different places.

The astonishing volume that some birds are capable of is in part because blowing through tiny holes can create big vibrations. However, the syrinx is also surrounded by an air sac, which acts as an amplifier.

But here's the really clever bit: a bird has pairs of flaps at the top of *both* of the Y-tubes, and can control the openings independently. It means one bird can make two completely different noises at the same time.

Mechanical noises

The syrinx isn't the only source of bird sounds; some birds also make noises using other parts of their bodies.

Beak

- **Beak-to-beak rattling**: In some birds, such as the Puffin, pairs face each other and clatter their beaks together in greeting, in a kind of avian nose-rub.

- **Beak snapping**: Several species make audible snapping noises with their beaks by closing them forcefully, especially owls but also birds as diverse as Pied Wagtails and Waxwings.

- **Drumming**: Great Spotted and Lesser Spotted Woodpeckers hammer a loud tattoo on tree trunks.

Wings and tail

- **Wind in the wings**: As birds fly, their beating wings create a rush of air, which can make an audible noise, especially so when the sounds are multiplied as a flock takes off. In a few species, however, the air moving over or through the feathers makes humming or 'singing' sounds. This includes the 'singing wings' of flying Mute Swans, Goldeneyes and Collared Doves (on take-off), and the 'thrumming' of Lapwing wings during its display flight.

- **Wing crack**: Several birds clap their wings above or below their bodies in flight, creating a loud whipcrack, including Woodpigeons, Stock Doves, Short-eared and Long-eared Owls, and Nightjars.

- **Wind in the tail feathers**: The Snipe has specially shaped outer tail feathers that create a throbbing hum in its display flight, which is called drumming.

Feet

- **Slapping**: Some birds, such as Coots and Mute Swans, slap and splash their webbed feet on the water to create a commotion.

Types of bird sound

Making sounds, and listening to sounds made by other bird species, is a major feature of most birds' lives. The language of birds may not be speech as we know it, but you have only to listen to birds for a short while to realise the complexity and range of their vocalisations, and how it is all designed to convey information.

To try and make sense of this language, we tend to categorise sounds according to their apparent meaning, and the obvious first step is to divide them into calls and songs.

- **Calls** tend to be short and straightforward, are given by both sexes and at any time of year, and often signal some kind of intent or action. Bird calls are also thought to be innate — in other words, a bird doesn't have to learn them and would be able to make them even if it had never met one of its own kind.

- **Songs**, meanwhile, tend to be longer and are mainly (but not always) given by the male, at a particular time of year, driven by hormonal changes. Bird songs are usually linked to defending a territory or attracting a mate, or both, but they have to be learnt at least to some extent. As we will see later, it is mainly the passerines, which we more commonly call 'songbirds', that have what we would describe as a song.

Track 3

The simple division of sounds into calls and songs is not perfect but holds true much of the time. Both categories can be further subdivided, so over the following pages we explore the different types, and what — with some poetic licence — they might mean.

Be aware, however, that hardly any species has a repertoire that includes each and every category of call. The range of sounds also varies according to the age, sex, and emotional and hormonal state of a bird, and some of these sounds might be used frequently, some very rarely indeed.

Different types of calls

- **Long-distance contact calls**: Many birds find themselves in a situation where they want to find members of their own kind, or locate mates and family members that are distant or out of sight. An effective way to do this is to have a rather loud call that says, 'I'm here; are you there?' Birds in flight often give this call to see if anyone replies from the ground below, or vice versa.

- **Short-distance contact calls**: Birds that are close together, whether on the ground or flying as a flock, will often make quieter noises merely to say to those around them, 'I'm still here; I'm feeling calm. There's no danger, and all is well.' These noises can sound somewhat conversational. In flying flocks, these calls may also be helpful for coordinating movements and are especially useful at night to keep contact in the darkness.

- **Anxiety and alarm calls**: Most birds need to take action many times a day to avoid danger. It is essential, therefore, to have a set of calls that send clear signals to those around them that range from 'I'm feeling rather nervous' through to 'Our lives are in imminent danger'. Such calls may also signal to the predator 'I've spotted you; there's no point chasing me'. They are calls we frequently hear because we are often the cause of them! Some alarm calls are danger-specific, such as the hawk call, which many different songbirds share and which clearly identifies that the danger is a fast-flying bird of prey. If you learn this call, you'll find more hawks and falcons as a result!

- **Excitement calls**: These calls tend to be louder and more intense than mere contact calls. They are typically made when there is no apparent source of danger, but something interesting is happening. Maybe a bird has chanced upon a tasty food supply, or found friends, or is sensing the time to migrate is approaching.

- **Mobbing calls**: These are a particular form of alarm call that signal when birds have found a roosting, predatory bird, especially an owl. They are a rallying call for other birds to come and harass the predator and to make its life so uncomfortable that it wants to leave.

- **Calls to do with take-off and landing**: Many birds signal their intent to take off or confirm they are doing so with distinctive calls. It ensures they don't get left behind and that the flock remains tight and coordinated. Some birds also have a landing call.

- **Calls to do with fighting**: When birds get into tiffs, maybe over a mate or food, it is useful to have a threat call that says 'Back off!' and hence resolves the situation without resorting to fisticuffs. However, if the threat doesn't work, the birds may have to move to full fighting calls. When one bird has won a contest, it may indulge in triumph calls, which is a feature of some swans and geese.

- **Calls to do with mating**: There are many calls to do with the intense moments between a pair, such as soliciting calls, calls immediately before and during copulation, and the celebratory calls afterwards.

- **Calls to do with parenting**: Many parents have specific calls when communicating with their offspring; some even talk to their unhatched eggs, encouraging the chicks to break free. Once the young have hatched, parents may have a call that says 'Stay close' or 'Freeze!'. A few wading birds lure predators away from their chicks by feigning injury, with special calls to complete the deception. In colonial species, such as seabirds and terns, there is enough nuance within the individual calls of parents and chicks that they can recognise each other, ensuring that the right chick gets fed.

- **Calls of young birds**: Many chicks have 'Feed me, feed me' calls, some of which can be quite insistent.

Different types of song

When a bird, such as a Blackbird, sits on a favoured perch in spring and repeatedly utters beautiful bursts of sound, it is clear to us that this is singing rather than calling – these bird sounds seem full of melody, similar to our songs. But what about something like a Tawny Owl hooting – is that a song? Or the *oooh* of a male Eider? Or the boom of a Bittern?

There is no absolute right or wrong answer to these questions. For the purposes of this book, however, sounds that birds use to attract a mate or defend their territory are referred to as follows:

- A **song** when it is given by a passerine (the songbirds – larks, warblers, tits, finches, buntings, etc.), even if the song is apparently rather tuneless. Some passerines also have a song flight in which each verse is often fuller, longer and more varied than their song when perched.

- An **advertising call** when it is given by non-passerines such as ducks, waders and gamebirds. And if it is given as part of a visual ritual, it is distinguished as a **display call**.

However, where a non-passerine has a particularly melodic advertising call, such as a Cuckoo or members of the pigeon family, most people would happily call these calls a song. This book, therefore, adopts a pragmatic approach – if it has the 'feel' of a song, it's called a song.

On page 15, we'll also look at how birds have to practise their songs. In some species, practising like this results in what are called subsongs and plastic songs.

Who is the intended target of the song?

It is the breeding season, and a Great Tit is sitting in a tree merrily singing *tee-cher tee-cher tee-cher*. Is he signalling to other males nearby that this piece of territory is occupied and if they try to muscle in, they will be repelled?

The male flies to another branch and sings a different verse, *wee-chee-chee wee-chee-chee*. Is he pretending to be a second male to make out that this territory is jam-packed with males? Or is he showing that he is such an accomplished singer that he must be fit and experienced?

And is a female listening for the male with the best choral skills and the most extensive repertoire? Or is she seeking out the best habitat, and then mating with the male who happens to have secured it?

Well, in Great Tits we know that their song is an auditory 'flag' for males to demonstrate their territories to each other and that females prefer males with lots of songs. However, in other species, the relative importance of the two functions is thought to vary. The Sedge Warbler song, for example, is almost entirely linked to attracting a mate, but when male Black Grouse (below) gather at their lek, a traditional sparring arena, their 'cat among the pigeons' display calls help determine who gets the best spot on the dance floor, and the females then choose the male in the prime position.

Subsongs and practice

You may think that birds emerge from the egg with their songs already hard-wired into them, ready to use at the start of the next breeding season. But for songbirds at least (the passerines), this appears not to be the case: song is something that must be learned.

The first lesson begins as soon as young males (and, in a few species, young females) hear the songs of the adult birds around them, which could be when they are still chicks in the nest. While at this stage they may not make any attempt to copy what they hear, the patterns of the songs apparently begin to lodge in the young birds' brains.

The process of then practising these patterns out loud goes through three stages that merge one into the other:

1. **Subsong**: Often in its first autumn but also into early spring of the next year, a young bird may sit, alone and self-absorbed, singing a rambling, introspective mumble with little structure and little resemblance to what it will ultimately become.

2. **Plastic song**: Then, during its first full spring, a bird develops its subsong to the point where it sounds more assured and the verses are more structured. However, at this stage the bird is still trying out different versions of its song.

3. **Crystallised song**: Finally, the bird settles on several verses that will form its set repertoire. In the Yellowhammer, one male can have as few as two verses that it uses, whereas a Robin may have more than a hundred. What we are now discovering, however, is that 'crystallisation' isn't as rigid as was once thought and that males of many species can continue to refine and augment their repertoire as they get older and more experienced.

Dialects

Just as people in different parts of the country have words, phrases and grammar that are peculiar to them, so populations of birds in a particular area can share the same verses and other sounds. These are known as dialects.

Sometimes these variations can be on a tiny geographical scale, such as in the Corn Bunting (right), where a population of 30–100 males sings its own unique song verses, different from those of neighbours barely a few miles away. The reason for this, in evolutionary terms, isn't clear, but it does appear that young female Corn Buntings may choose to mate with males singing a different dialect, helping to avoid inbreeding.

Dialects can also present on a larger geographical scale. Chaffinch 'rain calls' (page 76), for example – those monotonous 'lazy songs' of the male – may sound different in Scotland to those in the Midlands or southern England. So if you go abroad on your holiday, don't expect familiar birds always to sound the same as they do back home.

Imitation

When a Starling sings from your rooftop, it is easy to ignore its blips and whistles, pops and creaks. However, if you do stop and listen, you find that sometimes woven into its experimental electronics it will include quite astonishing imitations of other sounds.

It may be the call of a Buzzard or gull – birds that the male Starling might have heard flying over your house. However, he may include calls that he must have heard somewhere else, such as those of a Lapwing or Common Tern, or even sounds from the human environment – perhaps a mobile phone or even snatches of a human voice.

It is not only Starlings that mimic the sounds of other species: Wheatears can be fantastic at it; Jays frequently imitate Buzzards; Reed Warblers often start their songs with a few bursts of maybe Oystercatcher, Swallow or Bearded Tit; and I've even had a Robin in my garden repeatedly pull off an excellent Greenfinch impression.

Our best imitator is, sadly, incredibly rare in this country: the Marsh Warbler. One male can have a repertoire of the sounds of more than 80 different birds, including European species from its breeding grounds and African birds that it must have heard in the winter.

So why do birds imitate each other? It appears that the main reason is to show off! The females in some species love to find a mate with a vast repertoire, and what better way for him to increase the number of songs he can perform than by stealing some from other birds.

Duetting

A few bird species have a rather endearing quality in which the pairs sing or call together. The most usual form of duetting is where one bird starts and the other then follows, which is called antiphonal duetting.

Sadly, none of our birds is quite as accomplished as the Happy Wrens of Mexico are, which finish each other's sentences. However, we do have duetting geese, divers, grebes and White-tailed Eagles. My favourite, though, is the Crane: one bird will trumpet, sparking an instant reflex response in the other bird, such that if you didn't know otherwise it might sound like a single bird. It all helps create and reinforce a solid pair bond which, in a species like the Crane, is vital because each chick will need the support of both parents over many years to have any chance of surviving.

How to describe bird sounds

The building blocks of birdsong

The first thing we need to describe the sounds we hear is a consistent terminology.

The songs we sing are typically made up of a chain of notes that form clear sections: an introduction, verse, chorus and an ending. Bird songs and calls are not so narrowly defined, but it is possible to agree on some simple terms:

- **Song**: A bird's song consists of the full range of verses it sings. A very few species have a couple of song types, where one set of verses sounds very different from another. For example, the Wood Warbler has one song type that sounds like a spinning coin and another that sounds like a sweet *pyū pyū pyū*.

- **Verse**: This is one burst of song, with a defined start and end. Some species sing short verses, lasting little longer than a second; a few have very long verses, such as the continual outpouring of song from a Skylark. Some birds repeat their verses exactly; others sing a new verse each time.

- **Notes**: Each verse a bird sings is made up of a series of notes. For the purposes of this book, a note is not usually as simple as the single-pitched notes that make up most human music, but the note is still the basic unit in birdsong. So, for example, a Cuckoo sings two notes: *ku kū*

- **Phrase**: In some species, the verses are made up of distinct groups of notes called phrases. Sometimes these phrases are repeated. In the case of the Song Thrush, its song seems to be one repeated phrase after another with pauses in between.

- **Couplets/triplets**: This is a two-note or three-note repeated phrase, such as the *tee-cher tee-cher* of the Great Tit.

- **Syllable**: This is where each unit of pronunciation has one vowel sound. For example, the typical call of the Chiffchaff is a one-syllable *hweet*, while that of its close relative, the Willow Warbler, is a more disyllabic *hoo-eet*.

- **Repertoire**: This is the overall sum of all the different calls and verses that one individual bird or species makes.

Sounds and spelling

Are you wondering why, for example, the sound of the Cuckoo is written as *ku-kū*? This book uses simple pronunciation rules to give you a good idea of how to read transcribed bird sounds. Check out the guide on page 31, which sets out the simple pronunciation rules used throughout this book.

Sonograms

Later we will start to see how useful it can be when sounds are converted into images called sonograms. Sonograms are like a graph – time runs along the horizontal axis and pitch (page 22) on the vertical axis – conveying more complexity and precision in a few squiggles than a page of words ever can.

A true sonogram often looks full of smudges, which indicate the sound of recorded wind and other background noise. To help you recognise and understand bird songs and calls more clearly, we've created simplified sonograms like this:

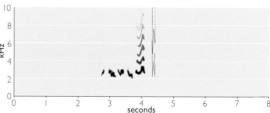

The seven attributes of sound

The next step is to describe what a bird song or call sounds like. Listen to Track 4, the song verse of a mystery bird, and try to describe it. Unless you are experienced, the likelihood is that you will find it quite tricky to put it into words. It is a kind of twittering, but can you say more than that?

The secret to describing bird sounds is that every sound a bird makes has the same set of attributes and you can focus on one attribute at a time. The seven attributes I find most helpful, and which we will return to again and again in this book, are:

1. **Duration**: How long does the sound last? Do the bird's song verses ramble on without it drawing breath or are they brief and succinct? Duration can also refer to the length of individual notes and how long the pauses are between the sounds.

2. **Pace**: Do the notes rattle along nineteen to the dozen or is it all laid-back or even plodding?

3. **Volume**: Are the sounds loud or quiet, the bird equivalent of shouting or whispering?

4. **Pitch**: How high or low is the sound? Is it like a bass, baritone or soprano? And does the pitch change at all, either from note to note or even within a note?

5. **Pattern**: Do the notes and their different durations, pace, volumes and pitch combine into a melody or rhythm?

6. **Timbre**: This is all about the character or quality of a sound and what makes one sound different from another even if they are the same pitch. We can use all sorts of adjectives to describe timbre, such as nasal, clear, piping, wheezing and trumpeting, but we can also try to transcribe the sound, turning the sounds as we hear them into words and letters.

7. **The overall effect**: When you put the first six attributes together, what is the sum of these parts, what is the overall effect of the sound? It may remind you of another sound, like a jangling set of keys, a steam train or a squeaky toy. It may also bring to mind a human emotion, such as sounding irritated or shy. A bird sound can even conjure visual images in the listener.

Breaking down sounds in this way helps us build a much more accurate picture. So, over the next few pages, we'll examine each of these attributes thoroughly.

Below is a sonogram of Track 4, the 'mystery bird' – play it again now and look at the sonogram at the same time. Don't worry if the sound and the sonogram still seem incredibly challenging to understand. As we examine the seven attributes, what you hear on the audio and can see on the sonograms will start to become clear.

Track 4

If you are unfamiliar with sonograms, you won't be used to listening to a sound while at the same time following the marks with your eye. However, it shouldn't take many repeats to begin to match each mark below with the song verse on Track 4.

You will find that closely spaced notes like this go by in a flash. You'll soon realise that birds often pack in lots of notes very quickly!

If you can read sheet music, you will have a head start when it comes to sonograms, for they have much in common. One difference is that in a sonogram, the horizontal axis is a precise ribbon of time.

Duration

Let's begin looking at the seven attributes of sound in more detail. Remember, the species by species guide is in Part 2, so, for now, don't worry about trying to identify the species in the recorded examples you listen to.

Track 5

The first attribute, duration, is all about how long each sound lasts. When you listen to a range of bird songs and calls, it quickly becomes clear how much they vary in length, whether that be individual notes or verses. These differences can form a useful part of your identification toolkit.

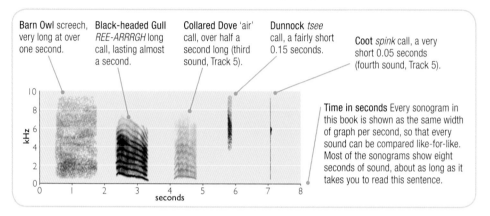

Barn Owl screech, very long at over one second.

Black-headed Gull *REE-ARRRGH* long call, lasting almost a second.

Collared Dove 'air' call, over half a second long (third sound, Track 5).

Dunnock *tsee* call, a fairly short 0.15 seconds.

Coot *spink* call, a very short 0.05 seconds (fourth sound, Track 5).

Time in seconds Every sonogram in this book is shown as the same width of graph per second, so that every sound can be compared like-for-like. Most of the sonograms show eight seconds of sound, about as long as it takes you to read this sentence.

Length of calls

Above are five bird calls shown on a sonogram. Don't worry about the vertical aspect – the pitch of the note – which we'll cover in more detail on pages 23–25. What you are looking for is the width from when it starts to when it finishes, which is how long it lasts (in seconds).

Very few single notes from any bird last longer than half a second, so something such as the Common Tern's second-long call really stands out, while the 1.5-second screech of a Barn Owl is exceptional.

Most single-note bird calls are much shorter, some a fraction of a second long. These are known as clipped or staccato notes. With experience, our ears are capable of detecting subtle differences between call lengths. For example, the notes in the *chi-chi-chit* call of redpolls could easily be confused with those in the *ti-di-dit* of the Linnet, but that 'ch' sound lasts a little longer, therefore on a sonogram it looks thicker. Try saying 'ti' and 'chi' out loud to hear the difference.

Sometimes, a bird's call is a combination of several notes so it can last several seconds, such as the *mwak mwak mwak* quacking of a female Mallard or the tittering whinny of a Little Grebe.

Length of song verses

Song verses and advertising calls are often much longer than most calls. Although one bird can sing verses of different lengths, most verses still fall within a typical range for that species, providing us with another clue to their maker's identity.

Although you don't need to go birding with a stopwatch in hand, it does help to become familiar with relative song lengths. So, for example, the Blue Tit's song verse rarely lasts 2 seconds so feels rather perfunctory. In contrast, the Wren's song verse rarely lasts less than 3 seconds and sometimes lasts as long as 8 seconds, so it can feel epic. However, that duration pales into insignificance compared with the reeling of the Grasshopper Warbler or Nightjar – they can sustain song verses for several minutes without a break.

Pace

Track 6

Whatever your musical tastes may be, we've all got an ear for gauging the speed of music we listen to. We know whether it is relaxed, like a slow waltz or a dreamy ballad, or whether it nips along at breakneck speed, such as the can-can or techno.

Having that ability means we are well placed to judge the relative speed of bird songs and calls. Here are snatches of five common bird songs that have different paces, as shown on a sonogram. Notice how the notes are more closely spaced on the timeline in faster-paced songs:

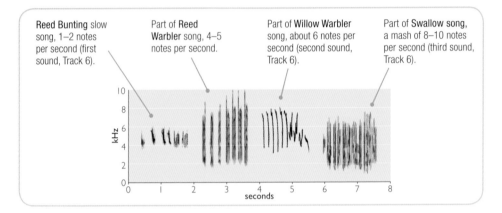

Reed Bunting slow song, 1–2 notes per second (first sound, Track 6).

Part of **Reed Warbler** song, 4–5 notes per second.

Part of **Willow Warbler** song, about 6 notes per second (second sound, Track 6).

Part of **Swallow** song, a mash of 8–10 notes per second (third sound, Track 6).

Repeat series, trills, rattles and buzzes

Many birds have songs or calls that repeat a single note in succession; in a large number of cases, the difference in pace of these notes is a significant clue to identification. You can judge how many notes a bird is making per second by trying to count them:

- **Up to eight notes per second**: You can count out loud up to about six in a single second (if you really go for it), and up to eight in your head. Bird calls at this pace are called a repeat series.

- **8–25 notes per second**: At this speed, it is too fast to count the notes, but we can still hear each of them. If musical in tone, this is called a trill; if somewhat percussive, it is a rattle.

- **More than 25 notes per second**: Sounds at this pace merge into a purr or buzz, such as the sounds produced by the Grasshopper Warbler (left) and the Nightjar (above).

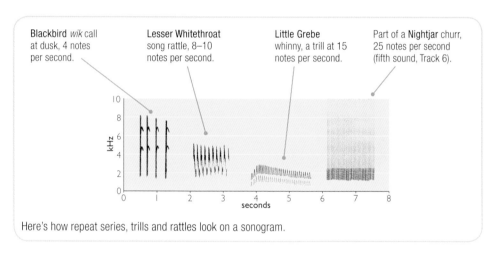

Blackbird *wik* call at dusk, 4 notes per second.

Lesser Whitethroat song rattle, 8–10 notes per second.

Little Grebe whinny, a trill at 15 notes per second.

Part of a **Nightjar** churr, 25 notes per second (fifth sound, Track 6).

Here's how repeat series, trills and rattles look on a sonogram.

Speeding up; slowing down

Some bird songs and calls alter pace as they go along. For example, the 'spinning coin' song of the Wood Warbler accelerates to its conclusion, as does the introduction to the Corn Bunting's 'shaken bunch of keys'. The best wind-up-and-release display call is probably that of the Dunlin, while the Red Grouse is notable in having one call that accelerates and another that slows down. But the master of control is the Nightingale; its *tyoo tyoo tyoo* verses build to a perfect climax.

What do birds hear?

Birds are able to process sound information about 10 times faster than we can. To them, humans must sound like we are speaking really slowly. And a Nightjar churring at 25 notes per second would sound more like we hear the slow *wik wik* of a Blackbird. It means that birds can discern all sorts of detail within their own songs, nuances that pass us by in a blur. This allows birds to recognise individuals of their own species when to us they can sound identical.

Volume

Just as some humans are louder than others, so are some bird species. Geese, gulls, Oystercatchers, Wrens, Nightingales and Cetti's Warblers are all at the big-mouth end of the spectrum. Mute Swans, Bullfinches, Shovelers, Pintails and Bar-tailed Godwits are some of the whisperers.

Track 7

The volume of a sound – technically known as its intensity – is measured in decibels. For the purposes of this book, however, a simple relative scale is all we need: are the bird's sounds very loud, loud, of average volume, quiet or very quiet?

Don't get confused between birds that make loud noises and birds that make lots of noise; volume is not the same as volubility. Long-tailed Tit calls, for example, are rather low in volume but the chatter is incessant; the Barn Owl is loud when it wants to be but most of the time is silent.

There are various factors that influence our perception of volume:

- **Distance**: Sound loses volume as it spreads out; it attenuates. So the relative loudness of a bird sound relies on you gauging what it would be like if you were next to the bird making it.

- **Pitch**: Low-pitched sounds travel much further than those that are high-pitched. This is why the deep boom of the Bittern carries so far across its reedbed home.

- **Weather**: Weather affects our perception of volume not only when the wind carries a sound towards or away from you, but also when cooler air in the atmosphere is trapped below warmer air, as often happens in winter and at sunset. In these conditions, sound can't escape upwards, making it seem louder.

- **Obstruction**: 'Things' get in the way of sound. For example, birdsong in the canopy of woodland gets soaked up by leaves and branches. Also, something making a loud noise will mask quieter sounds.

- **Hearing range**: Our ears are most attuned to sounds in the middle of our hearing range. This means that, to us, low sounds or high sounds don't seem as loud as medium-pitched sounds of the same intensity do.

- **Age**: As we get older, the sensitivity of our hearing tends to deteriorate.

Changes in volume

Another aspect of volume to listen for is how it changes throughout a call or verse. Does it come in with real punch, which is known as attack? Does it then sustain the volume or, indeed, increase it (crescendo), or does it fade away (decay)? The song of the Firecrest, for example, increases in volume over the course of each verse; Redstart song tends to start powerfully and then fade.

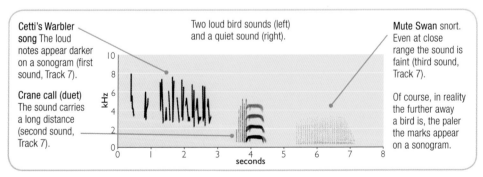

Cetti's Warbler song The loud notes appear darker on a sonogram (first sound, Track 7).

Crane call (duet) The sound carries a long distance (second sound, Track 7).

Two loud bird sounds (left) and a quiet sound (right).

Mute Swan snort. Even at close range the sound is faint (third sound, Track 7).

Of course, in reality the further away a bird is, the paler the marks appear on a sonogram.

Pitch

Pitch is how high – or low – we perceive a sound to be, and it is one of the most vital features when listening to and describing many bird sounds.

The pitch of a sound is determined, quite simply, by the number of times a sound wave vibrates each second; the faster the vibrations, the higher the sound. (At this point, scientists may well be having palpitations of their own, as technically 'frequency' describes the number of vibrations, and pitch describes how we hear it, but for the purposes of this book, our definition is fine.)

Pitch is measured in hertz (Hz). One sound wave vibration per second is 1Hz; 1,000 vibrations per second is 1,000Hz, generally written as 1 kilohertz (kHz). The human ear can typically hear sounds that vibrate in the range 20Hz–20,000Hz (20kHz). As we get older, we tend to lose the ability to hear sounds at the upper end of the range (or they will need to be louder before we can hear them).

Just to be clear, if things vibrate fewer than 20 times a second or more than 20,000 times, they still make a noise, but you'd need to be an elephant or a bat (respectively) to hear them!

The good news for us is that most bird sounds fall somewhere between 1 and 8kHz, well within the range for most of us; indeed, we are most sensitive to sounds around 2–5kHz, which is where songs of birds such as the Blackbird and Chaffinch are pitched. It means that most of us can become good judges of whether bird sounds are high or low.

Because pitch is shown as the vertical axis of a sonogram, we also get a brilliant visual representation of it. Here are some simple bird sounds at different pitches:

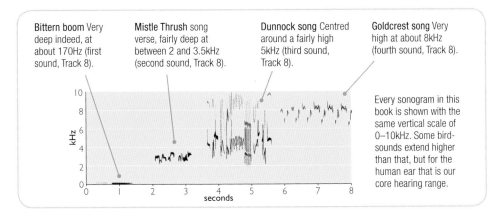

Bittern boom Very deep indeed, at about 170Hz (first sound, Track 8).

Mistle Thrush song verse, fairly deep at between 2 and 3.5kHz (second sound, Track 8).

Dunnock song Centred around a fairly high 5kHz (third sound, Track 8).

Goldcrest song Very high at about 8kHz (fourth sound, Track 8).

Every sonogram in this book is shown with the same vertical scale of 0–10kHz. Some bird-sounds extend higher than that, but for the human ear that is our core hearing range.

Pitch of complex sounds

So far, I've talked as if each sound wave is in one pitch. However, life is complicated, and most sound waves consist of sounds of different pitches piled on top of each other. But as long as one pitch is dominant, often the lowest one, that's the one we hear. This is known as the fundamental pitch. We'll explore more about the complexity of sounds in the section on Timbre (pages 28–30), including the fact that some sounds don't have a recognisable pitch, but for now we'll concentrate on sounds that have an obvious pitch.

Rising and falling pitch

How a sound changes in pitch is often what gives it real character and uniqueness. To write a bird sound in words and not show this aspect would be like trying to describe a song without mentioning the tune. Therefore, this book shapes how the text is printed to reveal the direction of pitch changes.

Curlew *courli* call. A strongly rising note (fifth sound, Track 8).

Buzzard *mew* call. A falling note (sixth sound, Track 8).

Eider *ooohhh* call. A gentle rise and fall.

Grey Plover call *plee-oo-wee*. A subtle but noticeable fall and rise.

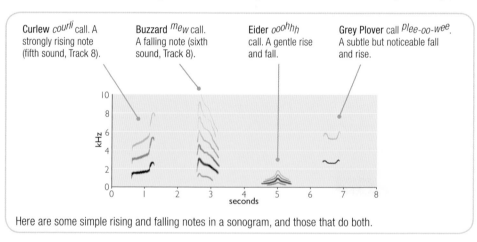

Here are some simple rising and falling notes in a sonogram, and those that do both.

Sometimes changes in pitch happen quickly. We can recognise these changes because we communicate using these kind of inflected notes all the time, and each can convey a distinct feeling. Let me explain:

- We use a **fast-rising note** when asking a question; try saying 'What?' with an upwards inflection to get the idea. It means that fast-rising bird notes also sound like a question.

- We use a **fast-falling note** when we give an answer in a definite way. Try saying 'Yes' as if you're answering a question categorically. Fast-falling bird sounds sound similarly finished, complete, firm.

- If fast-rising or fast-falling note changes happen really quickly, they sound **whipped**; if a sound actually jumps from one pitch to another, it is a **yodelled** note.

Some sounds go up *and* down, or vice versa:

- We give a **fast rise and fall** to express surprise, as in 'w^{o}w!', or to show we like something, as when tasting something nice, as in 'm^{m}m'. Lots of birds use this bend in pitch and sound pleasantly satisfied.

- We give a **fast fall and rise** when we question something we don't believe: '^Rea$_ll_y$?'

- Beware of the **Doppler effect**, which is where a sound from a bird coming towards you sounds higher-pitched than the same sound as the bird moves away from you. We often hear this with the 'singing wings' of Mute Swans as they fly past us.

- If these changes happen *really* fast and with a real leap in pitch, they sound '**whizzing**'.

- Vibrato is where a sound wavers rapidly and repeatedly back and forth; at a slower pace, the term used is **seesawing**.

Pitch can, of course, change repeatedly over the course of a verse of birdsong. This leads us neatly into the next section, which is all to do with pattern.

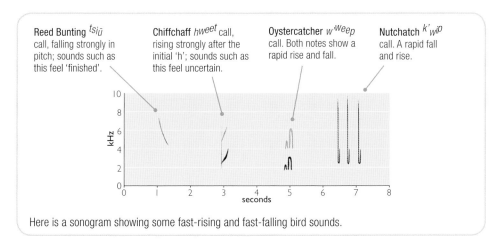

Reed Bunting *tsi$_{\bar{u}}$* call, falling strongly in pitch; sounds such as this feel 'finished'.

Chiffchaff *hw^{eet}* call, rising strongly after the initial 'h'; sounds such as this feel uncertain.

Oystercatcher *w$^{·wee}$p* call. Both notes show a rapid rise and fall.

Nutchatch *k'$_{wi}$p* call. A rapid fall and rise.

Here is a sonogram showing some fast-rising and fast-falling bird sounds.

Could you be tone-deaf?

About 4 per cent of the human population suffers from amusia, a condition in which they can't tell whether one sound is higher than another, often called 'tone-deafness'. If you are one of the unlucky few, this will make learning bird sounds more difficult. Search for 'amusia test' online to find out how good your pitch recognition is.

Pattern

We all enjoy a tune, and we love a rhythm. Rhythm is where a sequence of notes forms a pattern, either as it rises and falls in pitch or from the pace, duration and relative volume of its notes. Our ears are quick to pick up such patterns.

Track 9

Metronomic patterns

At its most simple, the notes within a call or song may stick to a regular beat, like that of a metronome or clock. The Chiffchaff's song, for example, is almost perfect in its 'one note at a time' rhythm, which can continue unbroken for 20 seconds or more.

In comparison, the Pied Flycatcher's song rhythm jogs along, not always perfectly in time but trying ever so hard to string its 5–10 notes into a steady series without fancy embellishments to the rhythm.

As we saw on pages 20–21, when a series of the same steady note speeds up, it turns into a fast series or even more rapid trill or buzz.

Multi-note patterns

Sometimes, instead of one steady series of notes, a bird will break up its sounds into blocks, creating the next level of rhythm. Take the Cuckoo: it sings *ku-kū*, then pauses, then repeats. What makes the call so pleasing to the ear is that the pause is usually about the same length as the *ku-kū*. To create this rhythm, steadily tap your finger in time with the dots or as you count the numbers:

•	•	•	•	•	•	•	•	•	•	•	•
ku	*kū*			*ku*	*kū*			*ku*	*kū*		
1	2	3	4	1	2	3	4	1	2	3	4

The Corncrake's two-note advertising call, on the other hand, has a slight pause between one 'phrase' and the next, so that each cycle is to the count of three (or almost!):

•	•	•	•	•	•	•	•	•
crex	*crex*		*crex*	*crex*		*crex*	*crex*	
1	2	3	1	2	3	1	2	3

The 'cycling' of three notes so that they seem to go around and around is a particular feature of many wader display calls, whether it be the slow, plaintive *per PEEE-yō* of the Golden Plover, the rolling *p'chū-ee p'chū-ee* of the Greenshank or the fast *t'lew-a t'lew-a t'lew-a* of the Ringed Plover.

There are longer cycles, too. The four-note rhythm of the Ptarmigan runs to that of 'Here comes the bride'; the five-note rhythm of the Woodpigeon is to the pattern of 'I don't want to go', where there is a strong stress on the 'don't'; and the Common Sandpiper's display call is about six notes repeating in pulsating cycles.

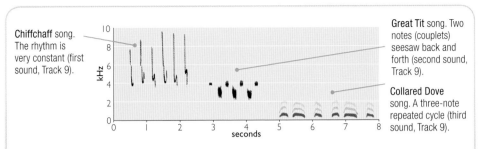

Chiffchaff song. The rhythm is very constant (first sound, Track 9).

Great Tit song. Two notes (couplets) seesaw back and forth (second sound, Track 9).

Collared Dove song. A three-note repeated cycle (third sound, Track 9).

Here are some simple rhythms on a sonogram – notice how the marks reveal the patterns. This can be a great way to remember bird sounds.

Complex patterns

The songs of many songbirds, plus some other notable vocalists, have patterns that are often much more complex than the cycling of a few notes.

In the Lapwing, for example, the long display call has a pattern that lasts more than four seconds and is split into three sections. Different males have slight variations on this.

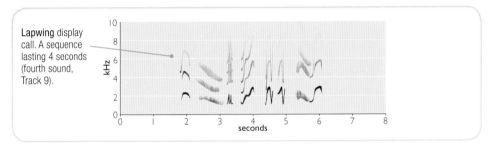

Lapwing display call. A sequence lasting 4 seconds (fourth sound, Track 9).

Complex patterns can be crucial to identification. For example, many people find the songs of Reed and Sedge Warblers impossible to separate, because both use scratchy and churring notes interspersed with sweeter whistles. The critical difference is not in the quality of their sounds but in their rhythms. The sonograms below reveal just how different their songs are – the Reed Warbler singing to an almost regular beat and the Sedge Warbler with its complex rhythms.

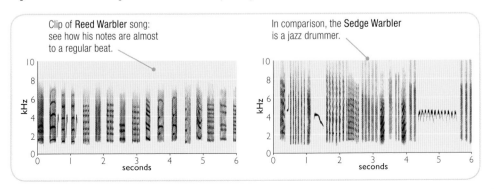

Clip of Reed Warbler song: see how his notes are almost to a regular beat.

In comparison, the Sedge Warbler is a jazz drummer.

Do verses repeat?

As well as listening for whether there is a pattern within a song verse, it is also useful to monitor whether a bird repeats the same verse each time, or whether it always sings something different. It is like a game of aural Snap, memorising the verse that has been sung and seeing if it matches the next.

Some species sing the same verse over and over again. However, they often have a handful of different verses in their repertoire, repeating one verse many times before switching to the next. In contrast, some birds sing a new verse each time.

Here are some examples of common species that are either 'repeaters' or 'changers':

• **Repeat the same verse**: Blue Tits, Great Tits, Chaffinches, Dunnocks, Wrens, Yellowhammers, Corn Buntings.

• **Change each verse**: Blackcaps, Whitethroats, Mistle Thrushes, Blackbirds, Robins, Nightingales.

Unusual patterns

A few species have songs and calls that exhibit atypical patterns. For example, Stonechats, Whinchats, Wheatears and the two Redstarts are prone to alternating between two very different calls with pauses in between. Hearing this can make the unwary think there are two different species present.

The most unusual song pattern is perhaps the pattern used by Song Thrush and the Ring Ouzel. Instead of singing a recognisable verse, these two species utter one or two notes, which they immediately repeat many times. They then pause and repeat a few completely different notes. The number of repetitions seems almost a matter of how much they liked the notes or not: it might repeat four or five times; it might not repeat at all!

Always remember that individual birds sometimes utter the strangest of sounds unlike anything else you may have heard from that species before.

Timbre

Play a note of the same pitch on a guitar, and a piano and they sound very different; the same is true for two people singing. The quality that gives each singer their own identity is sometimes called 'colour' or 'tone', but the correct term for it is timbre. And timbre is a crucial part of what makes different birds sound distinctive.

Track 10

Differences in timbre occur because most sound waves aren't a simple, regular vibration. They are influenced by subtle variations in the shape of the 'instrument', whether that's in the strings and resonating chambers of a guitar or piano, or a bird's syrinx, throat and mouth. The shape causes the sound waves to have all manner of little idiosyncrasies.

In fact, the timbre of a sound is further affected by the environment it is made in. Just as a guitar would sound different if it were played in your bathroom or your living room, so a bird's song in a wood on a foggy morning sounds different to the same song sung in an open field in the afternoon sunshine.

Of all the attributes of sound, timbre is perhaps the hardest to describe, and it relies on all sorts of subjective adjectives. Think of the human voice: a singer's timbre might be described as husky, pure, sweet, piercing, soft, breathy, melancholy… But, every effort to put timbre into words is worth it, and – as we will see – timbre also reveals itself in sonograms.

Pure notes, harmonics and overtones

One of the principal features that create timbre is that many sounds are made up of different pitches layered on top of each other, which are called overtones. In the section on Pitch (pages 23–25), we saw that our brain usually focusses on the fundamental pitch – the main frequency within a sound. But we still hear the other pitches as the 'colour' in the sound.

- If overtones are even multiples of the fundamental pitch frequency, they are called **harmonics**. They appear as widely spaced parallel lines on a sonogram, and they give a bugling or nasal quality to a sound. Our ears love harmonics; the best human singing voices are full of them.

- If the overtones are a random jumble, they are called **inharmonics**. On a sonogram, they look like a cloud or smudge of marks. We hear these sounds as noise rather than music, like the dry sound of a drum or a cough. In such sounds, there is often no recognisable pitch at all.

- **Pure notes** are sounds with no overtones which, in birds, are typically whistle sounds. These can be as beautiful and musical as notes with harmonics, but they lack depth and can often sound wistful rather than brassy.

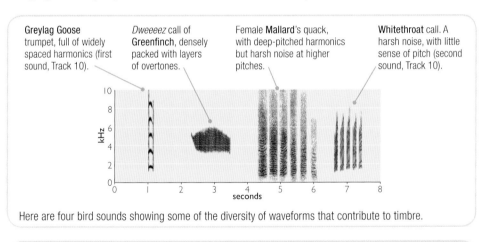

Greylag Goose trumpet, full of widely spaced harmonics (first sound, Track 10).

Dweeeez call of **Greenfinch**, densely packed with layers of overtones.

Female **Mallard**'s quack, with deep-pitched harmonics but harsh noise at higher pitches.

Whitethroat call. A harsh noise, with little sense of pitch (second sound, Track 10).

Here are four bird sounds showing some of the diversity of waveforms that contribute to timbre.

Key terms to describe timbre

- **Dry**: devoid of 's' or 'z' sounds
- **Fuzzy, smudgy, thick, sticky**: where notes are vague, ill-defined
- **Liquid, gurgling, bubbling, watery**: where notes flow freely
- **Musical, lilting, tuneful**: sounds that have the quality of human melodies
- **Nasal, bugling, trumpeting, brassy**: full of harmonics
- **Percussive, harsh, rasping, discordant, grating**: full of inharmonic overtones, lacking clear pitch

- **Pure, fluty, piping**: where notes are pure in pitch
- **Resonant**: echoing or prolonged sounds, as though made in a chamber
- **Rich**: full of layers of musical sound
- **Sibilant**: full of 's' sounds, hissing
- **Slurred, warped**: where one note bends in pitch to merge with the next
- **Staccato, clipped**: where notes are very short, abrupt and clearly defined
- **Thin**: usually a high-pitched sound lacking any deeper overtones

Putting timbre into words

When listening to bird sounds, we often try to make sense of them by thinking about them as if they were in our own language – as words. This is called phonetic transcription.

Track 11

There are some challenges with this approach. The first is the pace of bird sounds: many are so rapid that our brains can't keep up with them. The second challenge is that the same call can vary considerably depending on a bird's mood and urgency and the conditions at the time. And, finally, since birds don't speak English, transcribing their sounds is always approximate and subjective. It can, however, still be very instructive.

It is often easier to transcribe slow and simple sounds. Indeed, for birds such as the Chiffchaff and Cuckoo, we can transcribe each birds' distinctive sound closely enough that long ago their sound became their name, not only in English, but also in German where they are known as the *Zilpzalp* and *Kuckuck*, and in Dutch where they are the *Tjiftjaf* and *Koekkoek*.

Vowels and consonants

Transcribing vowel sounds is often reasonably accurate. The quack of a female Mallard is clearly not a *quick* or a *quock*, and the hooting of a Tawny Owl is based around a *hoooo* and not a *heeeee* or *haaaaa*.

Consonants are trickier, however. They include sounds that we make by blocking air and then letting it go with a burst (b, d, g, k, p and t), hissing and fizzing noises we make by pushing air through a small gap formed by our mouth (f, s, z, ch, sh and th) and the rolling 'r' sound that we make by trilling our tongue off the ridge at the top front of our mouth. We also make many consonant sounds with our lips, so it's no surprise that, when using a hard beak instead, many bird 'consonants' don't sound like ours.

And different people's transcription of consonants can vary incredibly. Is the Jay's harsh screech a *ksherr*, a *skaaach*, a *jarr* or a *kraih*? All of these are real examples transcribed by different authors. Maybe the Jay's screech is all of those things!

There is also the issue of pronunciation: if someone writes *kraih*, do they mean *cray*, *cry* or *cra-ee*? To try and avoid this potential source of confusion, we stick to simple pronunciation guidelines consistently throughout this book (see opposite).

How I hear bird sounds is almost certainly going to be different from how you do, but don't worry – my transcriptions will give you a helpful starting point and may help you to decide how you hear them. Phonetic transcriptions are an essential part of learning bird sounds but treat them as an indication of sounds and use them in combination with all the other tools.

Named after their call

Several British birds have such distinctive calls that they have come to be named after them. They include well-known examples such as Chough, Curlew, Hoopoe, Kittiwake and Twite, plus 'warbler', 'crow' and 'chat', and colloquial names such as Peewit (Lapwing) and Yaffle (Green Woodpecker). A few birds that might be less obviously named after their calls are the Coot, Jay, Rook, Teal, Turtle Dove and the Whimbrel.

Pronunciation guide

To deal with the vagaries of the English language, this book adopts the following consistent and straightforward pronunciation guidelines when transcribing bird sounds. Even when they refer to these guidelines, some people who have strong accents may still read the transcriptions in the book in different ways, but it should be close enough not to matter. It is crucial to note that transcriptions where a vowel has a bar over the top are long vowel sounds. I find it helpful to say the words in my head or as a whisper, rather than fully voicing them.

Vowel sounds

- **a** is short, as in *cat*

- **ā** is long, as in *car*

- **ai** is long, as in *air*, not *said*

- **aw** is long, as in *raw*

- **ay** is long, as in *say*

- **e** is short, as in *bed*

- **ee** is very long, as in *bee*

- **ew**, as in *new* or *view*, not *sew*

- **i** is short, as in *sit*

- **ī** is long, as in *light*

- **o** is short, as in *hot*

- **ō** is long, as in *open*

- **oo** is very long, as in *zoo*

- **ow**, as in *now*, not *low*

- **u** is short, as in *up*

- **ū** is long, as in *lute*

There are other vowel sounds in the English language but the ones above are perfectly adequate to cover most bird sounds.

Consonant sounds

- All consonant sounds are as generally understood, and none are intended to be silent

- A **k** is used rather than **c** to show that the sound is hard and not an **s** sound

- A **g** is hard, as in *gag* not *gentle*

- Extra vowels and consonants are added to lengthen the sounds, such as **brrrrr** or **ooooh**

CAPITALS signify a sound is loud (sometimes relative to the rest of the sound)

! at the end signify that the sound finishes abruptly

... signifies the sound continues in the same vein

- long   spaces   signify   long   pauses   between   notes

- shorter spaces signify short pauses between notes

- **hyphens-between-words** signify that notes run together

The overall effect

Although you won't find it listed in a science book as an attribute of sound, the overall effect of a sound can sometimes be the most valuable piece in your jigsaw. It is the impression the whole sound gives you, the images it sparks in your mind, the memories it evokes. We have so much personal experience of sounds to draw on that combines to trigger helpful connections.

For example, the calls of Waxwings sound rather like distant sleigh bells. This perception may be caused by a combination of timbre, pitch and pace but that barely matters because it is the evocation of Christmas and Santa Claus that is powerful and creates an abiding connection.

Here are other ways that bird sounds can create lasting connections in your mind. When you hear these sounds, think of the images, and they may well stick.

Machinery and objects

In a world full of non-natural sounds, it isn't surprising that some bird sounds sound somewhat similar:

- The advertising call of the Red-legged Partridge mirrors the chuffing rhythm of a steam train pulling out of a station, while male Egyptian Geese make the actual noise of the steam escaping.
- Mistle Thrushes and Ring Ouzels have rattling calls; the former sounds more like a football rattle, and the latter is reminiscent of a camera shutter.

- Bearded Tits ping like a miniature manual typewriter.
- Corn Buntings and Wood Warblers have accelerating metallic songs. The Corn Bunting is like the jangling of a bunch of keys, whereas the Wood Warbler is like a spinning coin.

Natural sounds

- The calls of the Sand Martin (below) are as dry as the sandbanks they burrow into.
- Robins sing the most liquid of all our bird songs, with all the trickles and gushes of a mountain stream, whereas the song of the Garden Warbler gives more of a sense of a full babbling brook.
- Some birds sound like other animals. For example, Great Black-backed Gulls bray like donkeys, whereas a ledge full of Guillemots sounds like a troop of monkeys. Birds that sound like small yapping dogs include the White-tailed Eagle and the Barnacle Goose.

Word-based mnemonics

For some birds, you can aid your memory by using English phrases that don't attempt to accurately transcribe the sounds but provide a memorable approximation. So while the Yellowhammer doesn't actually sing 'little bit of bread and no cheese' (or the polite version with a 'please' at the end), the sound has enough of the same pattern, duration and timbre to remind us of it.

Many people will have heard of the 'teacher teacher' of the Great Tit song, but why not also remember 'ouija ouija' to cement the slight difference of the Coal Tit song. Meanwhile, Stonechats call out for the 'sweet shop', Grey Partridges are afflicted with 'ear-yuk', and the Cetti's Warbler sings out for the 'chip, chip shop, chippy chip shop chippy chip shop'. I wish that more people could hear the sounds of the Leach's Petrel: 'I'm a little Leach's, and I'm in my hole' is the funniest little cartoon voice of any bird!

Linking bird sounds together

Sometimes, you can connect the sounds of very different birds, allowing you to store them mentally as 'bird sound partners'. For instance:

- Male Eiders, male Wigeon, Mediterranean Gulls and Little Owls all share whizzing calls that rise and fall in pitch.

- Swallows and Siskins both have a twittering song that throws in a completely out of character *dweeeez* towards the end.

- Lesser Whitethroats and Cirl Buntings both have a simple rattle.

- The Kingfisher *swee* call is like a stronger version of the Dunnock's *tzee*.

- Crested Tits and Turnstones both give fast ripple calls.

- And the unlikeliest of call pairs are the Common Sandpiper's *swee swee swee* and Black Guillemot's *swip swip swip*.

Human personalities

- We spend so much time surrounded by the noises of other people that of course we often hear little bits of human characteristics in the sounds that birds make. A call may seem shy, like the muffled whistles of Bullfinches. You might imagine Greylag Geese have their hands over their mouths. Listen for laughter, whether it be the guffaws of Cormorants or the embarrassed giggle of the Little Grebe (above). Some birds sound irritated or scolding, with Willow Tits and Peregrines the angriest of them all.

- The song of the Blackcap starts off self-consciously tentative but then it finds its voice, whereas the song of the Redstart starts confidently but then slightly loses its way.

- Many people struggle to tell the song of the Mistle Thrush from that of the Blackbird, but the former is the lovelorn poet, alone at the top of a tree, singing simple lines of rhyming poetry, whereas the latter is a much happier soloist that finishes almost every verse with a few bum high notes.

- Some bird sounds have the feel of a dance. The Chaffinch's is a bold, hopping jig, with a big 'ta-dah!' at the end. But the Willow Warbler's song sounds like a ballerina coming down a staircase.

- Meanwhile, Redwings hiccup, Dippers seem to play their songs backwards, and dare I mention flatulent Woodcocks?

Reading sonograms

The sonograms you have seen so far in Part 1 will have given you a good initial sense of how to read them, but here are two examples for you to check your understanding and consolidate your learning. With practice, you can begin to get a feel for a sound just by looking at the marks.

This sonogram is the bubbling display call of a **Curlew**.

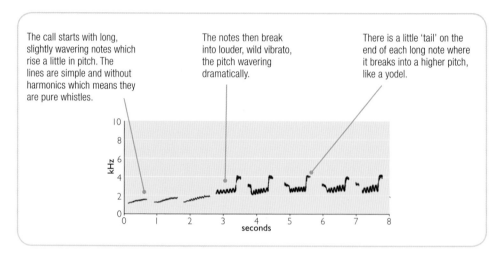

The call starts with long, slightly wavering notes which rise a little in pitch. The lines are simple and without harmonics which means they are pure whistles.

The notes then break into louder, wild vibrato, the pitch wavering dramatically.

There is a little 'tail' on the end of each long note where it breaks into a higher pitch, like a yodel.

This sonogram is of two song verses of a **Wheatear**, and the complete hotchpotch of markings is very typical of this most experimental of singers, which likes to sing two notes at the same time.

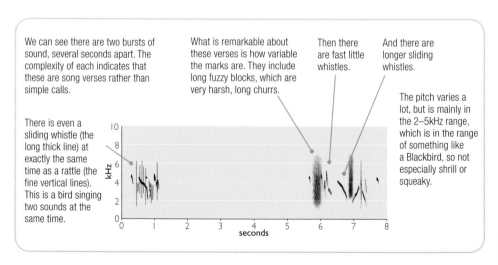

We can see there are two bursts of sound, several seconds apart. The complexity of each indicates that these are song verses rather than simple calls.

What is remarkable about these verses is how variable the marks are. They include long fuzzy blocks, which are very harsh, long churrs.

Then there are fast little whistles.

And there are longer sliding whistles.

There is even a sliding whistle (the long thick line) at exactly the same time as a rattle (the fine vertical lines). This is a bird singing two sounds at the same time.

The pitch varies a lot, but is mainly in the 2–5kHz range, which is in the range of something like a Blackbird, so not especially shrill or squeaky.

What sounds to expect: Where and when?

When you are out listening for birds, knowing what you are likely to hear is incredibly useful. You can substantially slim down the range of options by considering what time of year it is, where you are in the country and what habitat you are in. Doing this not only helps you make a quicker and better assessment of the possibilities but also has the effect of tuning you in, priming your radar with a set of anticipated sounds. The result is incredible, for an expectant ear is so often a successful ear.

Birds' distribution

While some species are widespread across the UK, many have restricted distributions. For example, there is no point hoping to hear Nuthatches in the far north or displaying skuas in the south, while Ireland has no Green Woodpeckers, Tawny Owls, Yellow Wagtails or Tree Pipits.

Weather impacts

Birds are much less likely to call and, in particular, sing or display in poor weather. They know that their sounds are much less effective when they are blurred by rain or swirled by a gale, and they must also get on with the business of staying warm and dry. So, a still and dry and sunny day is an invitation to sing their heart out.

North and west Scotland are especially rich in specialities, including exciting vocalists such as breeding **Red-throated** and **Black-throated Divers**, **Ptarmigans**, **Corncrakes**, **White-tailed Eagles**, **Dotterels** and **Crested Tits**. They are also the only places you are likely to hear displaying **Greenshanks** and **Whimbrels** and singing **Snow Buntings**.

If you want to hear **Puffins** guffawing, **Guillemots** gargling and **Choughs** chyow-ing, you have to head to remote sea cliffs and islands. In the case of **Gannets**, **Manx Shearwaters**, **Roseate Terns** and **Storm Petrels**, the choice is limited to a tiny number of sites.

The incredible song of the **Nightingale** is now heard solely towards the south and east of England. The southern half of the UK is also where you'll need to go to listen to the hyper-scratchy song of **Dartford Warblers**, explosive **Cetti's Warblers** and the simple crescendo shimmer of **Firecrests**.

To experience the slightly eerie ku-rrrr-loo of **Stone-curlews** at dusk, there are just two main centres of population, in central southern England and East Anglia.

The **Cirl Bunting** with its simple rattle song is almost entirely restricted to south Devon and Cornwall.

The circannual cycle

The flow of the seasons has a significant impact on which birds we hear, in part because most species are programmed to sing in the breeding season, but also because of the interchange of wintering, summering and passage birds. The chart below is a guide to some of the most prominent or evocative sounds that can be heard in each season, although, it will also depend on where you are in the country and what type of habitat you are in. There is no denying that bird sounds are a powerful part of nature's annual rhythm, and remain so today even in the middle of our cities.

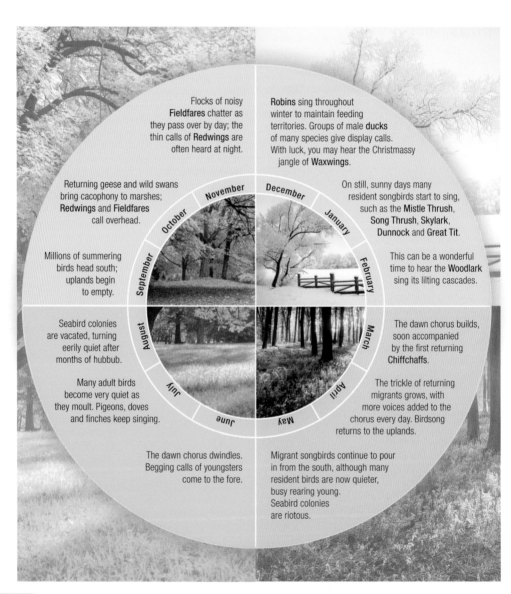

Flocks of noisy **Fieldfares** chatter as they pass over by day; the thin calls of **Redwings** are often heard at night.

Robins sing throughout winter to maintain feeding territories. Groups of male **ducks** of many species give display calls. With luck, you may hear the Christmassy jangle of **Waxwings**.

Returning geese and wild swans bring cacophony to marshes; **Redwings and Fieldfares** call overhead.

On still, sunny days many resident songbirds start to sing, such as the **Mistle Thrush, Song Thrush, Skylark, Dunnock and Great Tit.**

Millions of summering birds head south; uplands begin to empty.

This can be a wonderful time to hear the **Woodlark** sing its lilting cascades.

Seabird colonies are vacated, turning eerily quiet after months of hubbub.

The dawn chorus builds, soon accompanied by the first returning **Chiffchaffs.**

Many adult birds become very quiet as they moult. Pigeons, doves and finches keep singing.

The trickle of returning migrants grows, with more voices added to the chorus every day. Birdsong returns to the uplands.

The dawn chorus dwindles. Begging calls of youngsters come to the fore.

Migrant songbirds continue to pour in from the south, although many resident birds are now quieter, busy rearing young. Seabird colonies are riotous.

November — December — January — February — March — April — May — June — July — August — September — October

The circadian cycle

The 24-hour daily cycle brings its own rhythms of bird sounds. Of course, the exact timing of sunrise, sunset, dawn and dusk varies considerably depending on the time of year and your latitude. The sounds will also vary according to where we are in the circannual cycle and, not only that, weather conditions will play a large part on any particular day, with most birds far more likely to sing or call in still conditions.

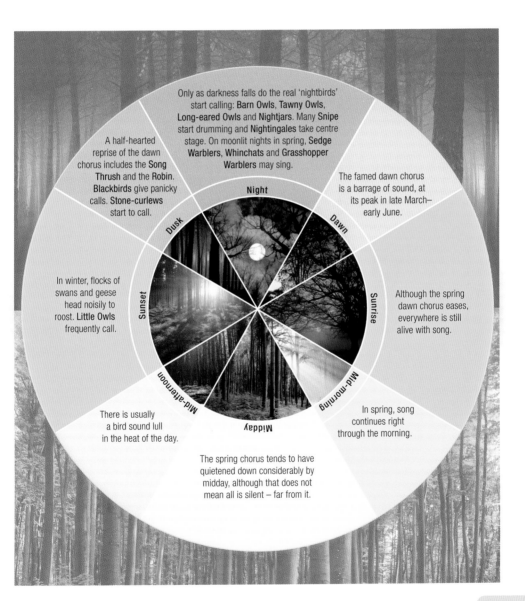

Only as darkness falls do the real 'nightbirds' start calling: **Barn Owls**, **Tawny Owls**, **Long-eared Owls** and **Nightjars**. Many **Snipe** start drumming and **Nightingales** take centre stage. On moonlit nights in spring, **Sedge Warblers**, **Whinchats** and **Grasshopper Warblers** may sing.

A half-hearted reprise of the dawn chorus includes the **Song Thrush** and the **Robin**. **Blackbirds** give panicky calls. **Stone-curlews** start to call.

The famed dawn chorus is a barrage of sound, at its peak in late March–early June.

In winter, flocks of swans and geese head noisily to roost. **Little Owls** frequently call.

Although the spring dawn chorus eases, everywhere is still alive with song.

There is usually a bird sound lull in the heat of the day.

In spring, song continues right through the morning.

The spring chorus tends to have quietened down considerably by midday, although that does not mean all is silent – far from it.

Night
Dusk
Dawn
Sunset
Sunrise
Mid-afternoon
Mid-morning
Midday

Habitats

Although you can't be entirely sure what birds you will find in a particular place, you can have a pretty good stab, based on the surroundings. The following pages highlight likely bird sounds in each main habitat.

Inner city

Even in the heart of our biggest cities, a few birds make their home and strive to be heard against the man-made din. It can make any birdsong you hear in the landscape all the more special.

Peregrines give peeved *rark-rark-rark* calls in spring above their **skyscraper** nests. Listen for screaming summer parties of **Swifts** high above.

Feral Pigeons sing their *kukker K'ROO* songs from **building ledges**. Woodpigeons are increasingly at home, too, adding their 'I don't want to go' song to the mix.

Large, loud birds add to the urban soundscape: **crows**, with their repeated *raarrk*; the rattling *schak-ak-ak* of **Magpies**; and raucous **Herring Gulls**.

Car parks are visited by **Pied Wagtails** with their *chizzik* flight calls and *swiz-zee* territorial calls. In a very few places, **Black Redstarts** sing their strange three-part song, the middle part sounding like someone walking on gravel.

It only takes an **avenue of trees** or a 'pocket park' to be able to hear the shimmery song of **Blue Tits**, seesawing **Great Tits** or mellifluous **Blackbirds**.

Although numbers have fallen dramatically, the blips and beeps of **Starlings** and the *chilp chilp* simple song of the **House Sparrow** may also still be heard.

Gardens and parks

A perfect place to hone your bird sound skills, this mosaic of green spaces gives many birds shelter and nesting opportunities. There is a network of green corridors for them to move along, plus homegrown seeds and berries, and generous hand-outs of extra food.

Jackdaws bring their perky *chak!* and *chyar* calls to any unguarded **chimney**. House Martins nest **under the eaves** from where they make their bubbling songs, while others spiral overhead, giving *preet* calls.

Collared Doves give bored *ku-kooo kū* songs from **TV aerials** and **telegraph posts**, and make breathy *a*^{*irrrrrrr*} calls as they land.

Taller trees offer song perches and nesting sites for **Greenfinches**, **Chaffinches** and **Goldfinches**, which bring their enthusiastic songs and varied calls to the suburban chorus.

Hedges provide corridors for constantly communicating **Long-tailed Tits**, and nesting sites for **Dunnocks**, which perch up to sing their simple shuttling ditties.

Large numbers of **Blackbirds** breed, the males sitting on **high perches** to illuminate the neighbourhood with their mellow songs. Sadly, there are fewer **Song Thrushes** these days, with their 'sing a phrase, repeat and move on' melodies.

There are plenty of **Blue and Great Tits**.

There are good, albeit decreased, numbers of **Starlings** and **House Sparrows**.

Ring-necked Parakeets occur mainly in the Greater London area.

There are many **Woodpigeons**, crows and **Magpies**.

Farmland

More than half of the UK's land area is farmland, with that figure approaching three-quarters in England and Northern Ireland. Profound changes in farming methods over the last century have affected much farmland birdlife from removing hedgerows and the use of pesticides to increased mechanisation.

Skylarks hang in the air high above **arable fields**, their rippling, extended verses sung from midwinter onwards. Listen, too, for the mewing of **Buzzards** circling overhead.

One of the noisiest spectacles is the massed cawing of **Rooks** and their young in a rookery, audible from afar.

Flocks of **Fieldfares** make noisy *tchak-tchak* calls in winter as they fly between **berry hedges** and **open pasture**, accompanied by quieter, 'hiccupping' **Redwings**. **Rough pasture** is also where **Kestrels** hover, but their chickering calls are only heard if they are breeding nearby.

In **spinneys** and **copses**, there is a tuneful spring chorus of **Chiffchaffs, Blackcaps, Robins, Wrens, Blackbirds, Song Thrushes, Chaffinches** and **Wood-pigeons**, interrupted by the *k'KOK!* of **Pheasants** and occasionally the chickering of **Sparrowhawks** and $p^{eeooooo}$ of **Red Kites**.

Where **stubble** or **game cover crops** are left standing in winter, flocks of finches such as **Chaffinches** and **Bramblings** gather, but also **Linnets** chorusing their *ti-dit* calls, and the tinkling charms of **Goldfinches**.

However, on a good many farms, helped by farmers' determined efforts to create space for nature, a wide range of birds thrive, and overall our farmed landscape supports a large proportion of our bird population. Some birds are heard almost nowhere else.

The 'steam train' calls of **Red-legged Partridges** and the *ear-yuks* of their much declined **Grey Partridge** cousins come from **field margins** and **crops**.

Swallows breed in **barns** and **outbuildings**, perching on **wires** to give their fast twittering song and darting around livestock giving upbeat *wit! wit!* calls as they go.

The deep *Whoo-er-wup* song of **Stock Doves** comes from **old barns** and **ancient hedgerow trees**, where Little Owls may give *hʷeeō* calls at dusk. The mix of **large trees** and **pasture** is ideal for yaffling **Green Woodpeckers**, and for **Mistle Thrushes** with their somewhat desolate song and football rattle calls.

Typical songsters of **spring hedgerows** include **Yellowhammers**, **Chaffinches**, **Dunnocks**, **Linnets** and **Whitethroats**, with, in a few places, the purr of the **Turtle Dove**, the rattle of the **Lesser Whitethroat**, *tett-tett* of the **Tree Sparrow** and 'key jangles' of the **Corn Bunting**.

There are plenty of **crows**, **Magpies** and **Jackdaws**.

House Sparrows, **Collared Doves** and **Pied Wagtails** find a home around the **farm buildings**.

Woodland

Woodlands are a changeable feast of bird sound: in springtime, they come alive with song, but in winter, large areas can seem deserted until you chance upon a roving party of Long-tailed Tits and their hangers-on when you can hear a flurry of their little conversations.

Where there are **mature trees**, especially **oaks**, the trunk-climbing specialists are much in evidence, the loudest calls being the fluid *kwip kwip* of **Nuthatches** and the sharp *kik!* of **Great Spotted Woodpeckers** as well as the latter's rapid tattoo drumming on **hollow branches**.

Much of the spring woodland chorus is made up of chiffchaffing **Chiffchaffs**, the faltering start and full-flow ending of **Blackcap** song and the bold 'skipping down the stairs' verses of **Chaffinches**. Blue Tits and Great Tits provide regular, repeated accompaniment.

Younger, more **scrubby areas of woodland** are the likeliest places to hear soft, lilting cascades of **Willow Warblers** and the babbling brook song of **Garden Warblers**.

Down in the **understorey**, **Wrens** are plentiful, but it can also be good for the *pitchū* sneezes and *bee-bee-bee* calls of **Marsh Tits**, and the liquid songs and *tip-tip* calls of **Robins**.

Easily missed are the high-pitched *seee* notes of **Treecreepers** plus their little, trilled song verses that end with a flourish. **Spotted Flycatchers** are also vocally understated, bar little *tsi chuk* calls.

The exact nature of the woodland and the age and diversity of the trees significantly affects the birdlife and hence the soundscape, with young woodlands supporting a very different cast to ancient woods, and conifers supporting yet another suite of species.

In **sheltered valleys** in mainly western and northern areas, the woodlands have their own distinctive chorus, with the spinning coin songs of **Wood Warblers**, the exuberant mixed trills of **Tree Pipits**, the simple rhythms of **Pied Flycatchers** and the 'bold-start, peter-out ending' of **Redstart** song.

Mature conifers hold a specialist suite of species, with **Coal Tits** prominent with their squeaky *WEE-jee WEE-jee* song. **Goldcrests** can be numerous, but their *sicily sicily* song is too high for some ears. The warped calls of **Siskins** are often everywhere, but **Crossbills** with their deep *chūp chūp* calls are unpredictable.

The harshest sound of the woodland is the sudden screech of the **Jay**: *SCARCH!*

As night falls, **Tawny Owls** begin to call, both the quavering hoots and the contact *ke-WIK!* calls of both sexes.

Other songbirds include **Blackbirds** and **Song Thrushes**.

Woodpigeons and **Pheasants** are often plentiful.

Buzzards and **Sparrowhawks** breed.

Woodcocks make roding flights over many woods.

Heathland

This rare lowland habitat is found on warm, sandy soils and windblown clifftops in southern Britain, where it is typically covered in heathers and gorse, with scattered pines and birches on inland sites, and peaty bogs in the hollows. In winter it can be desolate, with barely a bird sound unless you flush a Snipe from wet ground, 'sneezing' as it rises. Even in spring and summer, the species list is limited, but the compensation is some top-drawer songsters.

The male **Woodcock** telegraphs his wandering dusk flights (called 'roding') with a strange, intermittent *pizp!* and a snorting *warp!*

It is well after sunset before the **Nightjar** powers up his summer engine, an other-worldly churring, interspersed with wing clapping and *k'wip* calls.

The *chi-chi-chi   brrrrrr* song of **Lesser Redpolls** is often heard overhead in their bouncing song flight.

Either from **lone trees** or in the **skies** high above, male **Woodlarks** find endless ways to sing their gorgeous *lulu lulu* seesawing melodies down the scale.

Stonechats perch up giving 'sweet shop' calls and short verses, like scratchy snatches of **Dunnock** song. On some **southern heaths**, Dartford Warblers scold with nasal *tcharr* calls and sing in frenetic bursts.

You may hear **Meadow Pipits, Tree Pipits, Whitethroats, Yellowhammers, Linnets, Redstarts** and **Wrens**.

Crossbills may visit **mature pines**.

This is the prime site for the rare **Hobby** in summer, and occasional **Great Grey Shrike** in winter.

Rivers and streams

We all love to walk alongside a river, and rivers offer an excellent opportunity to hear many specialist species. There are different bird communities to find depending on the size and rate of flow of the river, and the vegetation along its banks, all enjoying the plentiful food and nesting sites it offers.

Kingfishers like **slow-flowing water** with **steep banks** for nesting, giving short *svee* and *svit* notes as they shoot by low over the water.

On **faster-flowing stretches** with **white water** and **exposed rocks**, the *zi-zit* of Grey Wagtails and *tzik* calls of Dippers are audible above the river noise. The latter also sings endearing verses – slow series of squelchy notes that sound like a record playing backwards.

Steep banks over the water are prime sites for colonies of **Sand Martins** with their super-dry *prrt prrt* buzz-calls. **Swallows** and **House Martins** often join them to hawk for insects over the water.

Exposed shingle beds are perfect for **Common Sandpipers**, which call *swee-swee-swee* as they whirr away low over the water. Listen, too, for loud *w-weeps* of **Oystercatchers** and, in a few places, the buzzing couplets of the **Little Ringed Plover's** display flight.

Where **bankside vegetation** grows densely, **Moorhens**, **Sedge** and **Reed Warblers**, and **Reed Buntings** take advantage.

Canada and **Greylag Geese**, **Mute Swans** and **Mallards** use many **slower-flowing rivers.**

Reedbed and fen

A unique bird community lives wherever there is permanently wet ground that is thick with vast stands of Common Reed. Once widespread across lowland Britain, this habitat is now primarily restricted to nature reserves, which at least give us ready access to its bird sound riches.

Many of the best sites for Nightingales are in **thickets** around **wetland margins**. Here, too, secretive **Cetti's Warblers** make their explosive outbursts.

In our **largest reedbeds**, Bitterns boom, while **Marsh Harriers** glide overhead, in early spring giving short $we^{e}o_o$ calls. On still days, you may hear the miniature manual-typewriter pings of **Bearded Tits**.

In **lowland areas**, Reed Warblers deliver their metronomic song. **Water Rails** clamber unseen, but you may hear their 'sharming', the special name given to their pig-like squeals.

Isolated bushes emerging through the reeds or rushes provide perfect song posts for **Sedge Warblers** and **Reed Buntings**, the former making what sounds like jazz-filled **Reed Warbler** song, the latter as though it's learning to count.

In winter, murmurations of **Starlings** shape-shift in the skies at sunset, although it is once in their reedbed roost that they actually murmur in sociable chatter. Other roosters may include **Pied Wagtails** and, in autumn, masses of **Swallows**.

You may hear **Wrens**, **Moorhens** and **Coots**.

In a very few places, **Cranes** breed in the **heart of the fen**, giving strident bugling calls.

Open water

Lakes, gravel pits and reservoirs can be busy places for birds, with a wide range of calls drifting across the water surface. The species present depend on the size and depth of the water body, and the amount of vegetation in the water and along the margins.

Grey Herons stalk the **water's edge**, but their frank calls are mostly heard in their heronries in **nearby trees**. The guttural barking of Cormorants also comes from their colonies.

On **open water**, Coot calls are prominent, a mix of sharp *cyūt* and *plik* sounds, while **Great Crested Grebes** growl. Above, elegant **Common Terns** give rasping *eeer-yarrr* calls.

Greylags and Canada Geese are noisy, **Greylags** having nasal *ungh-ungh* notes, and **Canadas** typically giving a rising *mmm-RUK!*

The whinnying titter of **Little Grebes** stands out, even if they themselves often stay close to cover. Listen, too, for the sudden *brook!* and *ki-keck* calls of Moorhens.

You need to be close to **Mute Swans** to hear their soft snorting. Much louder is the slapping of their feet on the water in territorial charges, and the throb of their wings in flight.

Female **Mallards** laugh heartily, *mwak mwak mwak*, louder than the frog croaks of the males. Listen, too, for the apologetic *bibs* from male Gadwalls, and the higher peeps of Teal. Male **Tufted Ducks** sound like cartoon mice quietly sniggering.

Open marshland and scrapes

On river and coastal plains – our levels, ings and washlands – shallow freshwater pools develop in winter in dips and hollows in the flat, open grasslands, and the ground is soft for probing beaks. The pools gradually dry out in summer, creating muddy margins. These 'scrapes' are often specially created on

Gulls visit the **shallow pools** to bathe throughout the year. **Black-headed Gulls** stay to breed, their unruly colonies are a din of raspy *RAAAARRRRGH* calls, while these days, you can increasingly expect a few cute $y^o w$ calls of **Mediterranean Gulls** among them.

Flocks of **Wigeon** graze the **flat grassy areas near the water** in winter, the whizzing $hw^{e}o_o$ call of the males a constant delight, and much better known than the females' very deep growls.

Where **bare mud** is exposed, wading birds can probe, with **Avocets** particularly noisy in the breeding season, a clamour of panicky *klūt* calls. **Black-tailed Godwits** are very rare breeders, but year-round flocks can give excited, seesawing ^{WI}CK-a ^{WI}CK-a calls.

Knowing the flight calls of passage waders really helps pick them out, especially the $^{t}y\bar{u}$ $^{t}y\bar{u}$ $^{t}y\bar{u}$ of **Greenshanks**, the *ship-ship-ship* of **Wood Sandpipers** and the cheerful sve^{ee} sve-sve^{et} of **Green Sandpipers**.

Perhaps the most evocative breeding bird of the **wet grasslands** is the **Lapwing**, the males displaying in a rollercoaster flight, their wings making a throbbing noise as they give their ^{peeeee} $^{i}d_{l}$-w^{it} i-w^{it} i-w^{it} pee-er-w^{it} calls.

nature reserves and can be some of the most exciting and diverse places for birdlife. This can make them full of sound, whether it is the calls of ducks and geese in winter or the display calls of waders and a cacophony of gulls in summer.

In winter, the **wildest coastal marshes** resound to the calls of thousands of geese, be that yapping **Barnacle Geese**, the deeper *brrup brrup* of **Brent Geese**, the squeaky *wil-a-wik* of **White-fronted Geese** or the double *ungh-ungh* and *wink-wink* notes of **Pink-footed Geese**.

On **extensive wetlands**, **Redshanks** breed, with lovely cycling display calls such as *tū-dlee tū-dlee*. **Snipe** pairs may also nest, switching between ground-based *chuk-a-chuk-a* calls and the amazing in-flight, tail-feather twanging display.

In eastern England and the Midlands, **where cattle graze**, you can hope to hear the *sweet* calls of **Yellow Wagtails**, which is usually as much as they can manage vocally. Expect some **Pied Wagtails**, too.

Greylag and **Canada Geese** add to the goose-fest.

Skylarks are common, filling the **skies above** with their endless songs.

Any **deeper water** is likely to contain dabbling ducks of several species in winter, but especially **Teals** and almost-silent **Shovelers**.

Estuaries and beaches

Where the tide flows in and out twice daily across sands, muds and shingles, masses of birds come to take advantage of the plentiful lugworms, shellfish and fish that live here, with their sounds drifting evocatively across the wild, open landscapes, changing with the seasons.

In winter, large flocks of **Brent Geese** visit larger **saltmarshes**, giving their deep *brr^up brr^up* chorus. **Wigeon** can also be common at this time, but any **Pintails** among them are only heard occasionally.

In the north, squadrons of **Eiders** sail up and down the **estuaries** and just **offshore**, sounding amusingly titillated with their *ahh^h-OO_OOH* calls.

On **sandy shores**, Sanderlings scurry at the **tide's edge** in winter, making sharp little *plik!* calls when disturbed.

Male **Shelducks** cheep like little ducklings as they excitedly pursue females in meandering spring flights, the latter giving accelerating *ag-ag-ag-ag…* calls.

On **shingle beaches**, Ringed Plovers breed, their bat-like display accompanied by *t'lew-a t'lew-a* calls. Large **tern** colonies use a few **protected beaches**, a hubbub full of the *kerrix* of **Sandwich Terns** and, in the north, the sharp clicking and *peearrrrr* calls of **Arctic Terns**.

Mudflats teem with wintering wading birds, including great flocks of **Dunlins** and **Knots**, but it is the bubbling **Curlews**, piping **Oystercatchers** and the ever-anxious-sounding *^tyū-pū-pū* of **Redshanks** that dominate the soundscape.

Cliffs and islands

Some of our most exciting bird spectacles are found where seabirds breed in vast colonies on precipitous cliffs along some of our most dramatic coastlines. Don't just focus on the visual treat or the smell! Soak up the sound atmosphere, which will last long in the memory.

In a few places in the west, **Choughs** rove the **short clifftop turf** giving their electric $c^{hy}ow!$ calls, while in the far north **White-tailed Eagles** may bark like terriers.

Kittiwakes calling out their names and the surging *ooh-ooh-ooh* and *rraar* of **Guillemots** dominate the cliff-face cacophony. **Fulmars** chip in with throaty clucks and cackles, and if **Gannets** are present, they have rolling *brrrrōk brrrrōk* notes.

Razorbills and **Shags** leave the din-making to the others, just making deep-pitched creaks, grunts and growls, but listen for the 'distant chainsaw' moaning of **Puffins** underground.

On a very few **islands**, summer nights are enlivened with the massed chicken-like cackles of **Manx Shearwaters** and strange, other-worldly purring noises among the rocks – **Storm Petrels**.

Rock Pipits flit around the **undercliff**, their song like that of the **Meadow Pipit** but more fizzy. You might also hear the winter ripple calls, *chi-ti-tik*, of **Turnstones** and the simple *kvit! kvit!* chatter of **Purple Sandpipers**.

Moors and mountains

As you head up onto higher ground, the trees give way to an open, rugged landscape of heather moors and craggy peaks, in places a vast and awe-inspiring landscape, and with its own sometimes sparse but evocative soundscape of bird calls and songs.

Male **Red Grouse** perch on mounds in the **heather**, giving their accelerating territorial clucking, or make short flights low across their territories, uttering a decelerating cackle that leads into a characteristic 'go back! go back!'

Sweet, lilting **Willow Warbler** song is omnipresent in spring and summer wherever there are scattered **birch** and **willow trees**. Nearby, **Whinchats** may sing their one-second, often three-section, ever-changing verses.

On **higher ground**, where the slopes turn rocky, **Wheatears** give their short but complex 'mashed-up' verses, which can include a lot of mimicry. Overhead, **Ravens** are the bass note of the **mountains**, a cronking bark.

Meadow Pipits are widespread in the **bracken** and **heather**, sometimes fluttering up in song flight before parachuting down, all the while giving their accelerating *sip sip sip…* songs. They also sit on the ground signalling alarm with repeated *t'rip* calls, for **Cuckoos** are often in attendance, either singing their name, or making their strange throttled sounds called gowking.

In only a few places, male **Black Grouse** gather at their traditional leks where they give their 'cat among the pigeons' display calls, a contrasting mix of sweet bubbling and fierce snarling.

Out on the open, windblown tops, birdsong becomes restricted to only a few species, but some of these are birds that are found almost nowhere else. However, come winter, there is often silence up here, bar sound of the the elements.

Ring Ouzels sing their sparing verses from **rocky gullies and steep ravines**, sounding like desolate **Song Thrushes**. Their varied hard calls include *tchok!*, which sounds like striking an icy pond with a rock.

Boggy areas are prime for curlewing **Curlews**, *peewit-ing* **Lapwings** and *chuk-a-chuk-a* or tail-drumming **Snipe**, with the chorus loudest at dusk and into the night. Male **Golden Plovers** circle in display overhead, giving lonesome *per PEEᴱᴱ⁻yō* calls.

On the **highest peaks and plateaus** of Scotland's upland you may chance upon some real mountain treats: the **Ptarmigan**'s call with a dry, ratchety frog croak, which when flushed is made to the rhythm of 'Here comes the bride'; **Dotterels** give simple *pwip* calls; and **Snow Buntings** have simple, sweet ditties given at a relaxed pace, plus their rippling main call, *tiririp*.

Grey Wagtails and **Dippers** use **larger mountain streams**.

Wrens are happy where there is **cover** and shelter, or vegetation and rocks.

Merlins may be seen but rarely heard, even less so for **Short-eared Owls**, **Hen Harriers** and – the most silent of all – **Golden Eagles**.

Putting it all together

You now have an almost complete toolkit to help you make sense of any bird sound you hear. You know that to describe a sound, you can focus on any or all of the seven different attributes of bird sounds: duration, pace, volume, pitch, pattern, timbre and the overall effect. You can also add into the equation what season and time of day it is, where you are in the country and what habitat you are in.

There is one last thing that will complete the picture when describing bird sounds and that is what the bird is actually doing when it makes the call. Is it flying, sitting, bobbing up and down? Or is it well hidden? These kind of details can be very pertinent.

You may recall that way back on page 18, we saw how a simple description of our mystery bird's sound as 'a kind of twittering' didn't tell us much about what the bird might be. Here's how we might describe it more fully using the seven attributes as well as details of place and time:

- It was a sunny day in May, in a bluebell wood in the Welsh mountains (**location** and **date**).

- There was a bird way up in the canopy that I couldn't see clearly at all, but it was repeatedly singing (**behaviour**).

- Each verse lasted at least 5 seconds (**duration**).

- It was rather loud, compared with the other birds around it (**volume**).

- The song had a clear pattern with several distinct sections – it had a mix of fast and slow trills, plus *weeee weeee weeee* notes, all linked together without a pause (**pace** and **pattern**).

- It was moderately high-pitched (**pitch**).

- It had a sweetness to it – the notes weren't at all scratchy or shrill (**timbre**).

- It was an uplifting song and full of energy as if the bird was really enjoying itself (**the overall effect**).

Hopefully, you can see that we've now got a much more complete picture of the singer and its song. It might be that you have a hunch what this particular bird is, in which case you can turn to the relevant species pages to check if the descriptions match. Or if you're not at all sure what the options are, you could use the habitat pages to review some of the likely species.

So what is our mystery bird on this occasion? While much of this description matches the Wren's song, the *weeee weeee weeee*, the lack of shrillness and the fact the mystery bird was up in the canopy should allow you to home in on the rapturous Tree Pipit (below).

Turning it into learning

While there's no denying that time spent outside among real bird sounds is an essential part of the process, true learning only really occurs when you nurture it. So, here are my Three P's for getting to grips with birdsong, with some ideas to make the process fun.

Practice

- **Tune in**: We tend to rely on our eyes for the bulk of our sensory information so, when you're out in nature, stand still, close your eyes and let your ears do the work. You'll be surprised how much more you hear.

- **Describe sounds for yourself**: Take out a little notepad, locate a bird singing and challenge yourself to find words to describe that sound. It really trains you to listen carefully.

- **Turn it into a game**: From this book, choose a species that you would like to hear, read the entry and then go out with the goal of finding it, listening to it and getting to know its songs and calls.

- **Take on a survey**: A great way of quickly improving your bird sound skills is to do a bird survey, many of which require you to note everything you see *and* hear.

- **Make recordings**: It's easy these days to record sound on your smartphone. And when you're ready to take it to a new level, check out how to make more professional recordings on pages 242–246.

- **Ticking them off**: If you like keeping lists of birds you see, why not also keep a list of birds you hear?

- **Learn from others**: Go on a guided walk with an expert, but make sure the guide and other birders allow you space to test out your skills. Or go out with friends and actively discuss what you all hear.

Patience

- **Stay positive**: Just as no one has ever become fluent in a foreign language in a day, so learning birdsong takes time. Consider every new sound you learn as a victory, rather than worrying if you have hundreds still to go. Don't be frustrated if some sounds don't sink in; give them a chance to embed. And don't berate yourself if you get something wrong – we're all on the learning curve somewhere, and none of us is at the pinnacle.

- **Take your time**: Don't feel you have to try and identify a bird sound instantly; give the bird some time to call or sing again so you can really get to grips with it.

Pleasure

But perhaps the best piece of advice I can give is to just enjoy it. Revel in bird sounds; be seduced by them; lose yourself in them. You'll learn so much more. And when you hear something astonishing, feel free to swoon!

Don't forget your field skills

Finding and appreciating bird sounds is all the easier if you use fields skills. At the very least:

- Don't make sudden movements.
- Don't head straight towards a bird – move at an oblique angle.
- Don't make lots of noise – leave noise-making to the birds.
- Keep white clothing concealed – otherwise birds will spot you a mile off and scarper.

PART 2
A Guide to Common Bird Songs and Calls

An introduction to our common bird songs and calls

This section of the book continues to work hand in hand with the audio recordings and will help you learn and understand the songs and calls of 65 bird species. It begins with birds we hear in towns and gardens, then moves on through woodland and farmland species, before finishing with some of the birds we hear in reedbeds and wetland edges.

The species in this section have been chosen because they are generally common and widespread, but also because they are birds whose songs and calls are often the best way of locating them. Indeed, many are species of woodland and the countryside that you are more likely to hear than see. In my experience, unless you are in particular habitats such as wetlands and islands, these bird sounds make up the vast majority of nature's symphony in Britain.

Get these sorted, and you will be able to quickly pick out anything you hear that sounds a little different.

The sonograms you'll see in this section have been created from specific audio tracks that come with this book so that you can follow with your eyes what your ears are hearing. I've also included written prompts to less common species that aren't covered in the recordings but whose sounds are either similar or provide useful comparisons. Remember that these recordings can't reveal all the variations that are possible in bird sounds, so in each case I have chosen examples that are the most representative.

As you work through this section, it is essential that you replay each track or phrase as often as you need to before moving on – repetition is an integral part of learning – and ten to fifteen minutes' practice at a time is plenty.

Track 15
Blackbird song

This is a great species with which to start improving your birdsong recognition. It is one of our commonest birds but with the most beautiful song that can brighten any spring day. It provides a benchmark for learning the songs of many other species.

Once a woodland bird, the Blackbird has adapted superbly to garden life. The males sing in gardens, parks and woods everywhere, although sometimes not starting until March.

Each male has a small number of favourite elevated perches from which to sing, often high in a tree or on a gable end or telegraph post. There, he will sit for many minutes at a time, returning several times each day from mid-spring onwards. He is most vocal at dawn and again in the evening.

Each verse lasts a rather short 2–4 seconds, and he then pauses for a few seconds before giving the next. Each male may have 30 or more different verses in its repertoire, and it will switch from one to the next every time, so each one sounds different.

It means you are listening for the overall feel rather than for any repeated melody. Each verse starts with a main phrase of five to seven notes, in quite a deep, warm baritone for a songbird. The timbre is melodious, the pacing relaxed. Listen how the individual sounds are more complex than simple, pure whistled notes, with little twiddles and turns, bits of vibrato and leaps in pitch. But beware – the further away you are, the simpler the song sounds.

There is a very useful extra clue. Most Blackbird verses end with an anticlimactic twiddle, very often high-pitched. It is rather quiet and so is difficult to hear at a distance, but if you are close enough it sounds almost like a mistake.

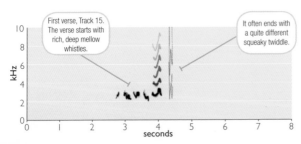

First verse, Track 15. The verse starts with rich, deep mellow whistles.

It often ends with a quite different squeaky twiddle.

Track 16
Robin song

Everyone recognises this perky bird with its orange breast and its confiding nature. However, its very variable song means that many people struggle to recognise it by sound. Think 'trickling water' to remember it.

Just like that of the Blackbird, the Robin's song is beautiful, with rather short, laid-back verses, most 1–3 seconds long followed by a pause before the next, and each is different, such that it can feel the song has no consistent structure. It is no wonder that the two songs can be easily confused when the birds are hidden in foliage.

The way to tell them apart is to focus on how the song *feels*. The Robin's song is higher in pitch, less mellow than the Blackbird's song. And there is a loose theme: each verse shifts back and forth between high- and low-pitched phrases, often starting high and ending low. But the thing to listen for more than anything is its overall liquid quality. By that, I mean its flow is like water issuing from a mountain spring, sometimes gurgling and trickling, sometimes with stiller, more languid moments.

In many places this is the first bird to sing in the morning at the very first hint of light in the east, and is the last to finish at night. It will often sing by streetlight, too, so much so that many people will imagine that a bird singing at such an hour must be a Nightingale.

The Robin is also interesting in that, after a quiet period in high summer, it resumes singing in autumn as it stakes out fiercely defended feeding territories, and both males and females can be heard singing sporadically through the winter. During these seasons, the verses tend to be slightly longer, softer and more melancholy.

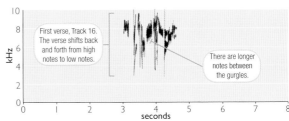

First verse, Track 16. The verse shifts back and forth from high notes to low notes.

There are longer notes between the gurgles.

Track 17: So, to recap…

Although both have ever-changing, short, laid-back song verses:

- **Blackbird** song is deep and mellow, with a terminal twiddle;
- **Robin** song is a trickling mountain stream with gurgles and stiller moments.

Track 18
Wren song

This tiny bird packs an absolute punch with its 'shrill with trills' song, which is far louder than seems possible for such a small bird and dominates many dawn choruses.

The Wren is our most abundant bird, found in overgrown gardens, woodland and any place where there is dense vegetation. It is rich brown with dark barring on its short wings and a perkily cocked tail, and is very mouse-like in its habits, disappearing into all manner of nooks and crevices in search of spiders and other invertebrates. Despite its diminutive stature, it is an incredible survivor, one of the few songbirds that are even found on remote islands.

The Wren usually sings from within cover, but sometimes from a little perch on top, belting out a few verses before darting off to another song post. Each verse is long, typically lasting 4–6 seconds, a rapid-fire bombardment of a hundred notes or more, delivered with gusto and volume. As the male sings, beak wide open, tail cocked, he turns his head repeatedly to ensure he is heard in all quarters. Many notes are in the very high 7–8kHz pitch range, which, together with the high volume, create the characteristic shrillness.

Each high-energy verse may sound like a blast of notes, but concentrate and you will hear that each verse links together about five to seven mini-trills and short series, with brief connecting notes between them. In particular, listen for a very fast, dry rattle or buzz in the middle of the verse, and a 'machine-gun' trill right before the end.

Each male has around five different verses in his repertoire; he repeats each one several times before moving on to the next. Neighbouring males will often have 'song duels'.

First verse, Track 18. Spot the three main trills joined by linking notes.

This is the ultra-fast machine-gun trill.

Track 19
Dunnock song

This modest little bird sings a rather hurried and anonymous ditty that is quite easy to overlook but is a common feature of many a garden and suburban soundscape.

The Dunnock is resident, found wherever there are plentiful hedges and bushes, avoiding either very open or well-wooded habitats. It looks like a fine-billed, grey-headed sparrow, and spends much time hopping around on the ground close to cover, constantly flicking its wings. It has an unusual *ménage à trois* (or *à quatre*) breeding strategy.

The Dunnock sings frequently in spring and even at times in autumn and winter, males perching within a bush or less frequently up on top. Each rather long verse is a fast, squeaky ditty, usually lasting at least 2 seconds and sometimes up to 4 seconds. It shuttles along at about eight notes per second, wandering up and down like a line of verbal scribble, without pause or change in pace – *diddly-diddly diddly-diddly*.

It varies between about 4 and 6kHz in pitch, without any mellow, deep notes, and so sounds a bit baby-voiced or squeaky. It doesn't have the Robin's gurgling notes or still moments, and so it feels very 'samey' all the way through.

The composition sounds so improvised that you might think each verse is totally different, but if you concentrate, you can detect that verses are often repeated almost exactly, with small changes often towards the end. Each male has a limited repertoire of perhaps six tunes, and other males locally may share elements of these, forming distinct dialects.

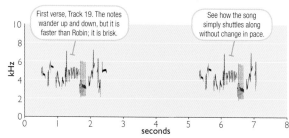

First verse, Track 19. The notes wander up and down, but it is faster than Robin; it is brisk.

See how the song simply shuttles along without change in pace.

Track 20: So, to recap…

- Wren song is long, loud and shrill, and packed with trills;
- Dunnocks sing *diddly-diddly* in shuttling samey verses of verbal scribble.

Track 21
Blackbird and Robin calls

Both sexes of these two common garden birds also give a range of calls throughout the year, most of them short single syllables, although sometimes repeated in long runs.

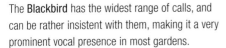

The **Blackbird** has the widest range of calls, and can be rather insistent with them, making it a very prominent vocal presence in most gardens.

Most arresting is the *wik* call, one of the staple sounds of dusk before going to roost, or when scolding a cat or owl. It is a loud *wik wik wik…*, rather high-pitched and sharp, and often given in long, uneven series, about four per second. It does

give the impression that the bird is deeply upset about something, the kind of noise that might make you urge it to 'calm down!'

Blackbirds also make a *chok* call, in short series of three to six calls at a rate of about six per second. In moments of what sounds like total panic, the *choks* accelerate quickly and rise wildly in pitch and volume before calming down again.

The **Robin** has a smaller repertoire of calls than the Blackbird, with its *tip* call being the most familiar. It is short and sharp, sometimes sounding more like *tik*. It is usually given either singly or more usually in stuttering series, with some rapid bursts of 10 or so thrown in – *tip tip tip-tip tipitipitip…*, almost like an irregularly dripping tap falling hard onto slate.

In alarm, it also gives a strong, piercing *seeee* call, almost half a second long, falling in pitch from a very high 9kHz starting point, and sometimes repeated time after time in slow series.

For a range of less frequent Blackbird and Robin calls, see pages 221 and 224.

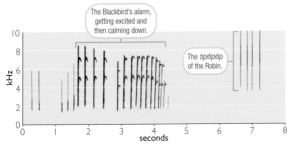

The Blackbird's alarm, getting excited and then calming down.

The *tipitipitip* of the Robin.

Track 22
Wren and Dunnock calls

To add to the *wiks*, *choks* and *tips* of the Blackbird and the Robin, the Wren and the Dunnock have their own short call notes.

The **Wren** (above) has two main calls, which – like its song – can seem quite loud for a small bird. It often makes them from deep within cover, betraying its presence when you might otherwise be oblivious to it.

The first is a simple *tjik* call, very easy to confuse with the call of the Robin. It is short and hard, like the sound of striking two flints against each other. It can be given singly, but is most distinctive when run into a very rapid series, with bursts of up to 10–12 notes per second that can sometimes run for 2–4 seconds.

The second call is similar, usually given when alarmed, but is more of a dry rattle, sounding like a clockwork toy

unwinding, *trrrrrrrrt*. It is given in pulses that can last anywhere up to 1 second, and at a fast pulse rate of about 30 notes per second. However, there are also shorter versions, such as *trit* – or very often *trit trit* – which sound like little double clicks of a clockwork mechanism.

The **Dunnock** (left) has one main call, which is quite different from the other three species, and is often the first indication of its presence, hiding within the undergrowth.

This simple, straight-pitched *tzee* or *tzzz* is rather weak and sibilant, like the sound of a squeaky hinge on a gate. It is short and unsophisticated, and fades in and out rather than being hard-hitting. It is often given in doubles, and sometimes turns into a little trill of four to six notes. When one Dunnock calls, other members of the breeding group often respond.

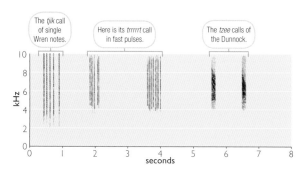

The *tjik* call of single Wren notes.

Here is its *trrrrrt* call in fast pulses.

The *tzee* calls of the Dunnock.

Track 23
Great Tit song

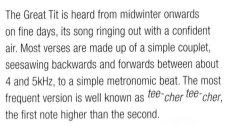

Strident and cheerful, the Great Tit's signature song is *tee-cher tee-cher*, but this is a bird with an extensive repertoire.

Bigger and bulkier than the Blue Tit, this is the largest of our tits but still a capable acrobat, usually seen singly or in pairs rather than large groups. It is readily identified by its black head with striking large white cheek patches and a thick, dark stripe down the middle of its bright yellow belly, but as with all tits it can be hard to get a clear view as it sings up in a tree.

The Great Tit is heard from midwinter onwards on fine days, its song ringing out with a confident air. Most verses are made up of a simple couplet, seesawing backwards and forwards between about 4 and 5kHz, to a simple metronomic beat. The most frequent version is well known as *tee-cher tee-cher*, the first note higher than the second.

He will repeat the 'teacher' couplet five or so times in each verse, then pause for a few seconds before repeating the same verse, doing this several more times. At dawn, expect the songs to be longer and the gaps shorter. However, each male has a few different verse types in his repertoire, each with a different rhythm. For example, one of the notes may be accented, such as *TEEE-cher TEEE-cher…*, or he may have a version that reverses the pitch change, *TEEE-cher TEEE-cher…*, or even moving to a triplet *tee-cher-cher tee-cher-cher….*

The timbre varies, too, from *tink-uh tink-uh…* to *tip s'lur tip s'lur…*; the versions you may hear from different Great Tits are almost endless, but all have a sunny, bold and confident feel, and a simple repeating pattern.

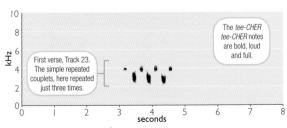

First verse, Track 23. The simple repeated couplets, here repeated just three times.

The *tee-CHER tee-CHER* notes are bold, loud and full.

The Blue Tit's song repertoire is almost as large as the Great Tit's, but includes a must-learn 'high intro dropping to shimmer' verse.

Everyone is so familiar with the Blue Tit at bird feeders and raising its broods in garden nestboxes, it is easy to be blasé about it. Take the time to look, however, and you are reminded what a beautiful bird this is, with its lemon underparts and blue tail, wings and crown.

Yet very few people recognise its song. It is certainly not as immediately attention-grabbing as the Great Tit's 'teacher teacher' song, and it also comes in a bewildering array of variations.

Singing from high in a tree, each male has about five different verse types within its repertoire, and even females are known to sing on occasion. Most verses are rather short, lasting 2 seconds or less, and all have a rather high, thin, sibilant feel.

However, the key to identification is the song's structure, for many verses start with two or three very high introductory notes (about 8kHz) and step down into a lower, straight-pitched, shimmering trill (around 5kHz). The most typical version is *sispi si-hi-hi-hi-hi*. Like the Great Tit, the Blue will repeat one verse several times before moving to the next in its repertoire.

However, be aware that some verses are a slower trill, such as *sip-sip zū zū zū zū*, or might yo-yo, such as *zip-zip zizizizizi zip-zip zizi-zip*. There are even simpler versions, too, without the trill, such as *tsi-tsi-sup-sup*, or a double version of this. However, notice that even these have that characteristic 'intro, drop to shimmer' pattern. Interestingly, Blue Tits are often prompted into sudden bursts of song on spotting a bird of prey, acting as a kind of alarm.

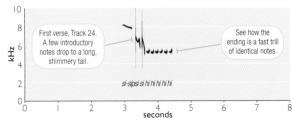

First verse, Track 24. A few introductory notes drop to a long, shimmery tail.

See how the ending is a fast trill of identical notes.

si-sipsi si hi hi hi hi hi

kHz

seconds

Track 25
Coal Tit song

From neck-craningly high up in a conifer tree, often well hidden from view, this dainty tit sings a song that makes it sound like the Great Tit's cute baby brother.

In looks, too, the Coal Tit is like a petite, stubby-tailed, monochrome Great Tit in tones of grey. However, the Coal Tit lacks the black chest stripe of its larger relative, but has a spreading black bib and a clearer white flash up the back of the head. The tiny bill is perfect for winkling out insects from between pine needles, and it is most common in

large conifer plantations, where its voice can be one of the defining sounds. Nevertheless, all it takes is a large hedge of Leylandii or a mature Scots Pine, Yew or Cedar in a garden or park to attract a pair.

The song follows the same basic pattern as that of the Great Tit, with a series of seesawing couplets or triplets, repeated about four to eight times. The key difference is the higher pitch, typically being between 5 and 8kHz, which, combined with the thinner, weedier timbre, makes it sound baby-voiced.

Another difference is that some verses are noticeably faster than the Great Tit's, up to five couplets a second. It does mean that the more familiar you are with the Great Tit song, the more the Coal Tit's cutesy and speeded-up version will stand out.

One very typical version is *WEE-jee WEE-jee…*, but as with the Great Tit each male has several verse types that it can switch between. Other typical examples include *chū-WEE chū-WEE …*, *WITZ-ū WITZ-ū…*, *chū-a-dee chū-a-dee…* and *tsū-wee tsū-wee…*.

Compare with similar calls of the Siskin, page 239.

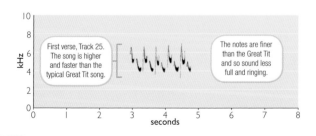

First verse, Track 25. The song is higher and faster than the typical Great Tit song.

The notes are finer than the Great Tit and so sound less full and ringing.

Track 26
Goldcrest song

This smallest of British birds has an ultra high-pitched song to match, beyond the hearing range of some people. Its cyclical pattern is the key.

Like the Coal Tit, the Goldcrest is found almost always where there are conifer and evergreen trees. Almost every churchyard will have a pair, and a single spruce or tall Leylandii in a garden will often host one, although it is in forest plantations that these birds are most prevalent. Many are sedentary, although most of those in upland woods relocate to lowland areas for the winter, and there is an additional influx from northern Europe in autumn, which stay for the winter.

Like the Coal Tit, males often sing from high up in conifer branches, and so are seen fleetingly or in silhouette. The song is fairly stereotyped in structure with subtle variations between different males. It is ultra high-pitched at around 7.5kHz. A verse starts by repeating what sounds like 'sicily' or 'silly-so' three to five times in a cycle, with a subtle but perceptible crescendo. It then finishes with a rather louder, slightly lower-pitched flourish: *si-si-lee si-si-lee si-si-lee sip-sip-sup*. In total, each verse lasts about 3–4 seconds.

The problem with Goldcrest song is that, as our hearing deteriorates when we get older, it is one of the first bird sounds that is lost to us. Many older people need to be fairly close to a singing Goldcrest to be able to hear it.

Compare with the Firecrest's song, page 218.

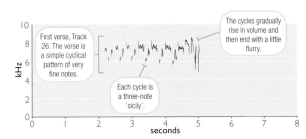

First verse, Track 26. The verse is a simple cyclical pattern of very fine notes.

Each cycle is a three-note 'sicily'.

The cycles gradually rise in volume and then end with a little flurry.

kHz / seconds

Track 27: So, to recap…

- Great Tits sing bold and ringing 'teacher teacher' couplets, plus endless variations of that;
- the short verses of Blue Tits are often high intro notes that drop to a shimmer;
- Coal Tits are like baby-voiced Great Tits, singing more *wee-jee wee-jee* than 'teacher teacher';
- the ultra-high *sicily sicily sicily* cycles of Goldcrests end with a flourish.

Track 28
Great Tit and Blue Tit calls

The tits and crests tend to be very conversational throughout the year, such that their sounds are often the first you know of their presence, especially in woodland.

The tits in particular have a wide vocabulary of calls, including persistent begging calls from recently fledged young in midsummer. Many of the quieter 'just staying in touch' contact calls, given as pairs and family groups move around, can be very similar across all three of our common species.

However, some calls stand out. In particular, the **Great Tit**'s *pink* call does exactly that – it is a cheerful, sharp *pink pink*, often indistinguishable from the *wink* of Chaffinch.

The Great Tit's alarm call, which it gives frequently, is a very simple, dry rattle of pitchless, rather deep, fuzzy notes, about 12 per second, like a mini-Magpie or shake of the maracas: *chrrrrrrrrt*. Some rattles are short, some longer and some introduced by a couple of higher notes. The rattle is usually steady and uniform, not wandering in pitch or pace.

The **Blue Tit** has a very similar alarm call but the notes within it are often faster, at about 15–20 notes per second, and it tends to rise in pitch. Like the Great Tit's alarm, it is often 'topped and tailed' with little calls.

Among its other calls, the Blue Tit's 'high-drop-to-low' theme, which is so prevalent in its song, is often heard, such as $^{ts\bar{\imath}\text{-}ts\bar{\imath}\text{-}}sup$ or $^{ts\bar{\imath}\text{-}}z\bar{\imath}\text{-}z\bar{\imath}$.

The Great Tit's alarm call, a rattling *chrrrrt*, here with an introductory *see* note.

The Blue Tit alarm call, like the Great Tit, but the series rises in pitch.

kHz

seconds

Track 29
Coal Tit and Goldcrest calls

The Coal Tit has distinctive, bendy high-pitched calls. Both the Coal Tit and Goldcrest have high-pitched *see-see* notes.

Many of the **Coal Tit's** calls have the same high pitch and distinctive timbre as its song, being rather cute and baby-like. In particular, they seem to be sounds that warp and bend in the middle. Typical versions include $si_{\bar{u}}$, $ti_{\bar{u}}$ or $ts_{\bar{u}}$. It is very typical to hear a sporadic conversation of the tiniest calls, which can sound like *tip* or *sip*, with a louder, deeper warped note interjected.

The Coal Tit (above) will also sometimes give thin *si* notes, often in short series, which is all the more confusing because this is the main call of the **Goldcrest** (right). In the latter species, the calls are given at varying strength, some as tiny as the bird itself but others of unexpected volume and sibilance. They are often at an incredibly high 8kHz in pitch but some of the quieter contact calls are even higher. It is very typical for the notes to be given in threes. They can run into a conversational, extended *sisisisi…* or can be longer, thin *seeeh* sounds.

So tit and crest calls are not always the easiest to assign to species, but listen for the more distinctive calls in their vocabularies and use any that you don't recognise to at least

guide you to where the birds are moving about in the canopy above. Usually, they will soon make one of their more recognisable calls to help you out.

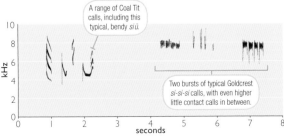

A range of Coal Tit calls, including this typical, bendy *siū.*

Two bursts of typical Goldcrest *si-si-si* calls, with even higher little contact calls in between.

Track 30
Woodpigeon song

The Woodpigeon is often the most frequent, and one of the staple, songsters of gardens and parks everywhere, with a deep pondersome cooing to a five-note rhythm that provides some of the bass notes of the garden chorus, year-round.

One of our most familiar birds, this hefty pigeon has boomed in numbers, and it is now as likely to be seen pecking around under bird feeders as it is out in woods and fields. The Woodpigeon is very vocal in spring, sometimes one of the first singers at dawn, and can be heard singing at almost any time of year; this is perhaps not surprising, as it has been found breeding in every month. Urban birds typically breed in spring, rural birds in summer.

The male usually gives its deep 'lowing' song from high in a tree, the sound emanating from deep within its chest, which inflates with each note. It is a simple, five-note cooing phrase – *ker KOOOO kooo ker koo* – repeated three to four times in succession. Imagine it saying, 'I don't want to go' in a rather weary way, the 'don'ts' becoming more insistent through the course of each verse. Listen closely and you will notice that the song actually starts 'Don't want to go' and finishes on an 'I' as if cut off prematurely.

Each five-note phrase lasts about 2.5 seconds, so that a full verse can last a long 10 seconds. The timbre is warm, the notes often having an attractive fuzziness rather than being pure coos.

When singing, the male barely opens its mouth, the sound actually being amplified by air sacs within his chest, and you do in fact get the sense that the sound emanates from deep within him.

For the Woodpigeon's two quieter calls and wing noises, see page 191.

Deep and lowing cycles of the song.

Don't want to go   I don't want to go   I don't want to go

kHz — 0, 2, 4, 6, 8, 10

seconds — 0, 1, 2, 3, 4, 5, 6, 7, 8

Track 31
Collared Dove song and 'air' call

The Collared Dove's sandy colouring betrays its exotic roots in dry, semi-desert lands, but it is now at home in many gardens, where its simple three-note song is pleasantly dull!

This species was perhaps the biggest bird success story of the twentieth century, when it spread westwards from its Middle East origins at an incredible rate to colonise most of Europe. It is a sedentary resident, most common where there are abundant sources of grain such as farmyards, and males are often to be found singing from the top of telegraph posts and rooftops.

The song is a consistent, formulaic, fairly deep three-note phrase, *ku KOOO kū*, one cycle lasting just over a second. It is often compared to the rhythm of a very bored football fan chanting 'U-ni-ted'. Unlike the Woodpigeon, where the phrase is always repeated a few times in succession, the Collared Dove will sometimes repeat its phrase many times in succession, with a short pause between each to form a full, extended verse, or it will sometimes give it singly with longer pauses in between.

There can be many variations, including where the three notes of the same phrase are all on the same pitch, or the first note can be slightly higher, like *ku KOOO kū*, or there can be a dip in pitch at the end of the long second note. Occasional variations include where the first note is slightly strangled – *kwo KOOO kū* – or where the overall timbre is muffled or croaky.

The Collared Dove also has a distinctive call – which I call the 'air call' – that it uses as it comes in to land. It is a simple, breathy, long *airrrrrrr*, usually given two or three times.

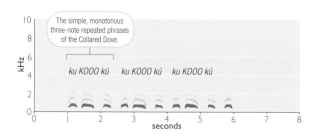

The simple, monotonous three-note repeated phrases of the Collared Dove.

ku KOOO kū   ku KOOO kū   ku KOOO kū

kHz

Track 32
Feral Pigeon song and nest call

So easily dismissed as it scuttles around your feet in city streets, the male Feral Pigeon actually gives a rather attractive bubbling display.

The Feral or Town Pigeon is derived from a native species, the Rock Dove, which just about hangs on in the wild on remote coasts of the far north and west. Domesticated for centuries, escaped birds and wayward 'homing pigeons' now live a feral existence on the 'cliffs' of the urban world, and come in a wide range of plumages from all white to almost black. As well as in city centres, Feral Pigeons are also found around farmyards, arable crops and cliffs.

Although fairly vocal throughout much of the year, their calls can get rather lost in the urban hubbub.

Males woo the females by raising their iridescent neck feathers and excitedly strutting, bowing and pirouetting directly in front of them, while giving a *kukker K'ROO* call from deep in their chest air sacs. The first notes are staccato, the ending like a muffled bubbling.

Males also sit near their nest site and give a series of half a dozen or so deep moaning, groaning *OOOOOH* or *OOOOO-uh* calls. Each note is more than half a second long, with a prominent, long first syllable, rising slightly in pitch then dipping to a weak rather than clipped ending.

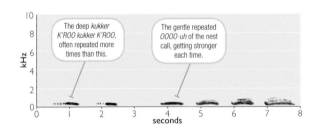

The deep *kukker K'ROO kukker K'ROO*, often repeated more times than this.

The gentle repeated *OOOO-uh* of the nest call, getting stronger each time.

Track 33: So, to recap…

- Woodpigeons moan 'Don't want to go, I don't want to go, I don't want to go, I';
- Collared Doves give a bored chant, 'u–NI-ted u-NI-ted', and call 'airrrrrrr' as they land;
- Feral Pigeons either bubble *kukker K'ROO*, or gently groan *OOOOO-uh* in series.

Loud, persistent and confident, the key-to-learn repertoire from one of our commonest species includes the male's oft-repeated song, which is a bold jig down the stairs with a theatrical ending.

The Chaffinch is a familiar sight at many bird tables or pecking around on the ground beneath. Males are attractive, grey-blue above and off-salmon below, while females are almost as dowdy as a female House Sparrow, best told by the pale flashes in the dark wings.

Males set up territory wherever there are mature trees, including in many parks and larger gardens, and are then very vocal in the breeding season, usually singing from high within a tree, although often not fully exposed on top.

The song is neat, bright and brash. Each verse is rather short and concise, lasting about 2–3 seconds. The notes drop down the scale sequentially, on each step giving a short, quick series of two to six repeated notes. At the end, he rounds it off with a flourish, such as *chew-eeoo*, often rising in pitch. So bring to mind the image of an entertainer on stage, gaudily attired and wearing rather heavy boots, doing a bold jig down a staircase. He does a few quick steps on each level before finishing at the bottom with a theatrical 'ta-dah!', arms outstretched to get his applause.

He then pauses for a couple of seconds, and then repeats exactly the same verse. This can go on for several minutes from the same perch, before he swaps to another of the handful of verse types in his repertoire, and repeats that one instead.

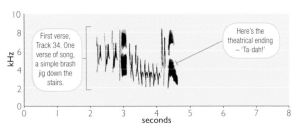

First verse, Track 34. One verse of song, a simple brash jig down the stairs.

Here's the theatrical ending – 'Ta-dah!'

kHz

seconds

Track 35
Chaffinch calls

The Chaffinch also has three primary calls, which – once learnt – provide a great basis for then picking out something different among their flocks.

An interesting call that can be attention grabbing is the 'rain call'. The male Chaffinch gives it in spring and summer, and it is indeed alleged to foretell impending rain. It may actually be a primitive song that the bird uses in less than perfect weather conditions or when he doesn't feel he needs to do a full-throttle jig.

It is one note or sometimes a double note uttered repeatedly, tediously, about once a second, for many minutes. The exact note used varies from place to place, and one male can switch between different notes, but frequent examples include a simple *hweet*, a reedy *zreep* and a double-note *wi-jit*. It can really stand out for its monotony, such that you wonder what that bird is that seems to be stuck in a rut.

Both sexes also give a strong, bold, clear *wink!* call, sometimes sounding like *chink!*, primarily in the breeding season, often repeated although not with the endlessness of the rain call. It is very difficult to tell from the Great Tit's cheerful *pink!*

The Chaffinch's main flight call is a soft *tyūp*, used to keep in contact with other birds around it, and is often given as a flock takes flight. This is the call you will hear from flocks in winter.

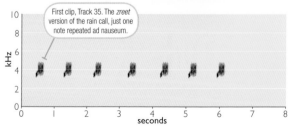

First clip, Track 35. The *zreet* version of the rain call, just one note repeated ad nauseum.

The key to learning this attractive finch's lively song is its tinkling 'tickle it' call, which it interweaves with its vivacious trills.

Gregarious and feisty, the adult Goldfinch has a sharply defined blood-red face and pure yellow flashes in the wing. Young birds, however, have plain, rather pale grey heads and streaky underparts, but share the yellow wing flashes.

Once taken by the thousand for the Victorian cagebird trade, the Goldfinch is now a much increased and very familiar visitor at garden bird feeders, even in the middle of cities, and it breeds in loose colonies.

The frequently heard call is a run of incredibly short notes which bounce about all over the place, and are best remembered as the 'tickle it' call. They give a characteristic tinkling quality: *tik-a-lit   tik tik tik-a   tik-a-lik-a-lit*. Listen, too, for quieter, high *sik* notes, at about 6kHz, dotted in between.

The male often sings from an elevated perch, sometimes in a circling song flight, and especially before roosting. In autumn, he will gather with his buddies to sing communally in treetops, creating quite a volume. The song is incredibly variable and can sound like a formless, tinkling twittering, but listen for those all-important 'tickle its', linking together fast, bright trills

and connecting longer notes. It can all flow without pause, and a verse can either be short or extended to 10 seconds or more.

Listen for some other notable calls, too. Often when disturbed, Goldfinches will give repeated, longer, pitch-bending warped calls, such as $z^{re}e$ or $dsee_{oo}$. Also, when two or more birds face off over food, they give loud, fast *chrrt* $ch^{rr}t$ threat calls, stretching forward aggressively, clearly livid with each other!

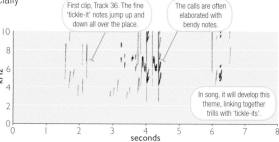

First clip, Track 36. The fine 'tickle-it' notes jump up and down all over the place.

The calls are often elaborated with bendy notes.

In song, it will develop this theme, linking together trills with 'tickle-its'.

Track 37
Greenfinch calls and song

As with the Goldfinch, recognising the Greenfinch's gentle calls helps you to learn its song. This bird also often sings a very long, wheezy note.

Males live up to their name, being moss-green all over with yellow flashes along the wings and sides of the tail, while females are much dowdier but still with the yellow flash along the wing. This was once a common garden finch, but numbers have crashed since 2005 when they became susceptible to the respiratory disease trichomonosis.

It has a range of rather attractive calls, the main call being a little trill of rather sweet ringing notes, *ji-ji-ji-…*, *jū-jū-jū…* or *dibbidib…*. There is nothing harsh about it, just velvety simplicity. The notes can be given as doubles, triples or up to eight in each trill.

In song, the male sits high in a tree or launches into a distinctive, slow-flapping display flight, meandering in wide circles above his territory like a bat, singing all the while. It is a joyous extension of his calls, a series of melodic mini-trills, each usually separated by a short pause. Once every few trills, he gives a long, slow wheeze song – *dweeeeez* – almost a second long, sometimes rising in pitch but often falling. In fact, the song can be no more than the wheeze repeated every few seconds.

In anxiety or alarm call, Greenfinches typically give a one-, two- or three-syllable-long whistle, with a very strong rise in pitch or rise-fall-rise, such as *dweee*, *dwee-ū* or *dwee-ū-ee*.

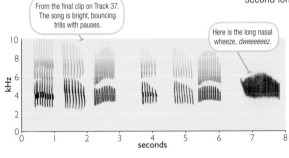

From the final clip on Track 37. The song is bright, bouncing trills with pauses.

Here is the long nasal wheeze, *dweeeeeez*.

kHz

seconds

Track 38: So, to recap…

- **Chaffinch** song is a bold jig down the stairs ending with a theatrical 'ta-dah!';
- the 'tickle it' calls of **Goldfinches** are woven into its excited tinkling songs;
- **Greenfinches** mix up runs of gentle *jū-jū-jū* and *dibbidib* trills with long *dweeeez* notes.

Track 39
House Sparrow call and song

Although much declined, the House Sparrow is still widespread, giving us plenty of opportunities to hear its cheerful if primitive song.

This is a bird that seems to spend its life picking up the crumbs at the table of human life. The population crash towards the end of the twentieth century is still poorly understood, but it seems not enough chicks survive their first year. Encouragingly, there are early signs of recovery, especially in Scotland, Wales and Northern Ireland.

This is another bird for which its main call is the way to start learning its sounds. Both sexes chat with each other using variations on a single chirping note, *chi*^{il}*p*. It is bright, short and confident, but see if you can hear two half-syllables running very quickly together: *chil'p*. You can see this clearly in the sonogram, and to our ears it sounds like 'the needle jumps' a little.

This then forms the basis of the male's rudimentary song, which it gives from perches near its nest site, whether that be a branch, the top of a bush or a gutter edge. It is little more than a slow, uneven series of the *chilp* call and subtle variations, one to two notes per second in random order, such as: *chi*^{il}*p chow*p *sh*^{eel}*p chi*^{il}*p...*. At his most excited, he droops his wings and half cocks his tail.

Listen too for 'Sparrow chapels' in which flocks of House Sparrows gather in the heart of favourite dense bushes to chat among themselves, a hubbub of their main calls plus soft, deep, muttered *churp* and *cheep* sounds, mixed with the long threat call, *chrrrrrrrrrp*.

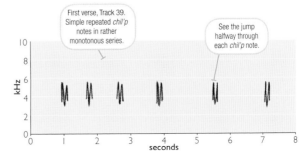

First verse, Track 39. Simple repeated *chil'p* notes in rather monotonous series.

See the jump halfway through each *chil'p* note.

kHz

seconds

Track 40
Starling song

Once omnipresent in both urban areas and the countryside, the Starling is now absent from many parts of the UK, but its incredible song can still be heard in many neighbourhoods, full of whistles and clicks and laced with all manner of clever imitations.

The Starling is gregarious and feisty, and its plumage is – as its name suggests – covered in stars. As in the House Sparrow, numbers have crashed disastrously and they continue to decline in both our breeding population and the hordes that arrive from the continent in autumn. In a few places, they still gather in dramatic numbers in winter, forming incredible shape-shifting manoeuvres in the evening sky before going to roost.

Resident males sing mainly between March and June, choosing prominent perches such as rooftops and television aerials. At the song's most intense, a male will droop his wings and flicker them or wave them in a sculling motion, his glossy throat feathers flared.

The song is rather quiet but each extended verse, up to a minute long, is an incredible jumble of electric squiggles, blips, pops and chuckles, like an old radio being tuned in. It often starts with a long, whizzing, tremulous whistle – *weeeeeee*. Very typically, clicking noises underpin parts of the song and are made simultaneously with the other sounds. There are often times in a verse when it is as if the 'record gets stuck'.

However, the fun thing to listen for is an experienced male who will throw in all sorts of imitations, often of other birds but also non-avian sounds, such as machinery, phones and frogs!

Starlings will also sing in communal chorus when flocks gather by day in trees. When alarmed, the chorus can cut dead in an instant, which is called a dread.

For the Starling's wide repertoire of calls, see page 221.

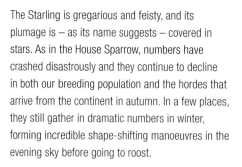

Starling song is often a complete hotchpotch.

See how little clicks run in the background under other noises.

Many verses include these long sliding notes, *weeeeeee*.

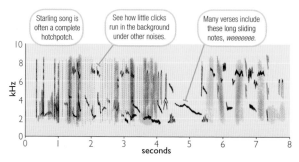

Track 41
House Martin call and song

This summer visitor spends much of its time high in the sky, where it gives its *preet preet* contact call. If you find an active nest, you can enjoy the delightfully frothy song.

Only now are we beginning to get a sense of where the House Martin spends the winter, somewhere high above the central African rainforest. Some people confuse these birds with Swallows; they are similarly dark above and white below, but they have a much shallower fork to the tail and the tell-tale plumage feature is their square, white rump patch. Returning birds mainly arrive from mid-April and depart by early October.

Their half-cup mud nests, with a small entrance hole in one corner, are built right up under the eaves on the outside of a building with a clear flight path in, rather than in a barn or outbuilding; a few still breed on their original cliff haunts. Otherwise,

they spend their time catching insects, often high in the sky but coming down low over the water when the weather is poor.

The main flight call is a rather lisping *preet* or two-syllable *pritit* or *chirrit*. The calls are particularly intense when trying to lure youngsters from the nest. Conversational twittering can continue in the nest during the night.

The full song is given especially at the nest itself, and is a summer treat. It is a rather cute, mushy babbling of uninterrupted ultra-fast notes that can last many seconds. Listen for the *preet* note, which is woven in from time to time, but the song also returns repeatedly to much deeper notes. Overall, the song has a sense of 'wetness', of frothiness – imagine an endlessly bubbling pan of soup!

For the House Martin's alarm call, see page 211.

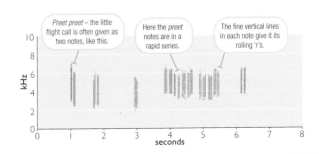

Preet preet – the little flight call is often given as two notes, like this.

Here the *preet* notes are in a rapid series.

The fine vertical lines in each note give it its rolling 'r's.

Track 42
Swift calls

The excited screaming of Swifts over the rooftops is the evocative natural soundtrack of evenings in high summer in those towns and villages still fortunate enough to have them.

This scimitar-winged, aerial master arrives back in UK airspace in early May and most have gone again by early August. Recent satellite tagging has proved that young birds remain airborne for the first two years of their life, and this is a species that can sleep and even mate on the wing. However, numbers are falling, thought in part to be due to nest holes being destroyed as houses are renovated; much conservation work is underway to safeguard known sites and to install external nestboxes and, in particular, special integral nest bricks.

The Swift is fairly vocal, especially in the evening but mainly around the nest sites and not usually when feeding or migrating. To describe a bird sound as a scream doesn't make it sound particularly appealing, but the Swift's main flight call is an exuberant, excited scream – *zreeeee(t)!* It is high-pitched at around 6kHz, each call usually about half a second long. The calls can be rather clear in timbre, or with an audible buzz, and if you listen closely many finish with a slight drop in pitch at the clipped ending.

The calls are usually delivered by small groups of up to 20 birds, either when zooming in daredevil flights low through the streets or wheeling in tight knots at considerable altitude. Together, the calls create a joy-filled chorus of fast-moving sound.

For the Swift's wing noises, see page 196.

High pitched, each screamed note is rather long.

See the fast, buzzing vibrato.

zreeeeeh

kHz

10
8
6
4
2
0

0   1   2   3   4   5   6   7   8
seconds

Track 43
Pied Wagtail calls

The most numerous and widespread of the two urban wagtail species is the Pied, a bird often seen in car parks, quiet roadside gutters and playing fields, skipping perkily over the short turf and tarmac, where it often gives its cheerful calls.

This is a familiar bird for many people, and well named, for it is indeed strongly pied, with a black back, cap and bib, and white face, forehead, belly and wing stripes, and it has a habit of merrily wagging its moderately long tail. Juveniles are greyer but still monochrome.

'Chiswick flyover' is a great way to remember its sharp *chizzik* flight call, which it uses regularly throughout the year, especially on take-off and in flight. The call is usually given in two syllables but sometimes the bird adds an extra – *chiz-iz-ick*. Each call is usually given on every dip of its deeply undulating flight course, at a rate of about one a second. As the word 'chizzik' implies, the two syllables are clearly different sounds, and the second is often slightly lower in pitch; also, each subsequent call tends to vary a little in its pronunciation.

The Pied Wagtail also has a territorial call, again heard throughout much of the year and not with the full function of a song. It is given most often from the ground or a perch. Like the flight call, it is also usually two syllables but is more musical, being a slurred *swiz-zee*, *shwer-zee* or *che-wee*.

In winter, large numbers will gather for the night, often on supermarket roofs, in reedbeds and even in city centre trees, and they will make these sweet notes when arriving.

For the Pied Wagtail's little-heard song, see page 231.

For the Pied Wagtail's little-heard song, see page 231.

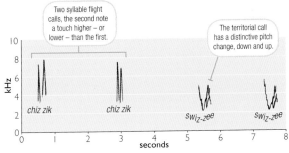

Two syllable flight calls, the second note a touch higher – or lower – than the first.

The territorial call has a distinctive pitch change, down and up.

chiz zik *chiz zik* *swiz-zee* *swiz-zee*

Track 44
Grey Wagtail calls

This, the longest-tailed of our wagtails, is most at home on mountain streams, but in winter it is increasingly found in towns and cities where it makes itself at home in the urban roofscape and parks. Therefore, knowing its piercing call is useful.

Seen well, the Grey Wagtail is a delightful bird, even in winter when it is grey above and with a splash of bright yellow under its tail. In breeding plumage, the male in particular flushes with richer yellow underneath, and he develops an attractive black bib and white moustache. However, it is the Grey Wagtail's ultra-long tail that stands out, both when the bird is flying and when it lands and wags it in an exaggerated manner.

Its flight call is given frequently, especially when landing or taking off, and is very similar to the Pied Wagtail's *chizzik*. Most often it is a double note, but it is sometimes given as triples or just a single. This is a call designed to be heard above the background noise of fast-flowing water, and so it is harder, sharper and more penetrating than the Pied Wagtail's call, sounding like *zit-zit* or *zi-zit*.

The two notes are more clearly separated than the run-together *chizzik* of Pied Wagtail, and they are nigh on identical, compared with the Pied's two notes, which usually differ slightly from each other in pitch.

For the Grey Wagtail's mechanical song and other calls, see page 231.

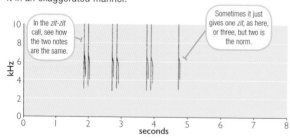

In the *zit-zit* call, see how the two notes are the same.

Sometimes it just gives one *zit*, as here, or three, but two is the norm.

Track 45: So, to recap…

- The **House Sparrow**'s song and calls are cheerful but rudimentary *chilps* and *chowps*;
- the **Starling**'s song is quiet, all electronic clicks, blips and *wheees*, with amazing mimicry if you're lucky enough to hear it;
- **House Martins** have a lisping *preet* call;
- the **Swift**'s daredevil screams are an excitement-filled *zreeee*;
- remember the **Pied Wagtail**'s flight call as the 'Chiswick flyover', but listen too for its *shwer-zee* call;
- and listen for the same note twice, *zit-zit*, in the **Grey Wagtail**'s call.

Track 46
Carrion Crow and Hooded Crow calls

Where you are in the country determines which of the two crow species you will see. Despite looking so different, there is little to tell between their harsh calls.

Visually, the **Carrion Crow** (above) is all black and easy to confuse with both the pale-billed Rook and larger, wedge-tailed Raven, whereas 'Hoodies' are easy to identify in their neat grey 'waistcoats'. The two are rarely seen together because they occupy different parts of the country: the **Hooded** (right) is a bird of north-west Scotland, the Isle of Man and Ireland, while the Carrion is found across the rest of the country. Neither species tends to be seen in tight flocks, although several dozen can congregate in major feeding areas, such as out on coastal town beaches and in some urban parks.

Both crow species are fairly vocal, especially in the breeding season, and they will also call in flight. Most often, however, they will call from a perch high in a tree.

The word 'crow', as you might guess, is in reference to their calls. However, the usual sound is a harsh, loud, deep *rarrrk* or *kaarr*, noticeably dropping in pitch. It is typically repeated three or four times, about one call a second, in a laboured, mechanical way, the bird leaning forwards and then thrusting its head upwards with each note in a rather aggressive posturing. It can sound malevolent, and has

much of the function of a song: it is very much a sound that says 'I'm master here'.

Listen, too, for variations, sometimes rather flatter in pitch, sometimes more gargled or strangled, or lacking bass pitch and so sounding weedier.

For the crows' raptor alarm call and the rarely heard song, see page 205.

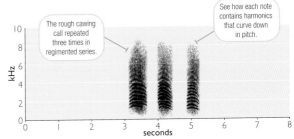

The rough cawing call repeated three times in regimented series.

See how each note contains harmonics that curve down in pitch.

Track 47
Jackdaw calls

This sociable small crow can't seem to help but make its distinctive, short and snappy calls wherever it goes. It usually sounds happy or excited, and in fact you'd be forgiven for thinking that the 'jack' in 'Jackdaw' is derived from the sound.

There is something quite endearing about the Jackdaw. Smaller than a Carrion Crow, it has cuter proportions, a grey-tinged head and twinkling silver eyes. It also has a sociable nature that includes deep devotion to its lifelong mate: they fly and feed together, and often snuggle up, preening each other. Several pairs can breed within loose colonies, using cliffs, quarries, ruins and churches, but for many of us, we are most familiar with them when they choose to build their nests inside our chimneys.

The Jackdaw is very vocal throughout the year, especially in the breeding season when several pairs may gather at dawn and dusk for an excited aerial sprint race, dashing in wide circles low over the rooftops, calling as they go. Very large numbers also gather in winter to roost in conifer woods, and they can sound equally excited before bed and then as they head back out into the fields at dawn. The calls help these sociable birds maintain their bonds.

Their sounds are high-pitched compared with most of those of crows and the Rook, with three staple calls making up much of their loud repertoire. Two are very short and snappy, *chyak!* and *chyok!*, the third a rather longer *chyar!* Even saying the sounds reveals how percussive and staccato they are and, indeed, rather fun! They are often given in little outbursts, inspiring others around them to join in.

For deeper calls and alarms, see page 204.

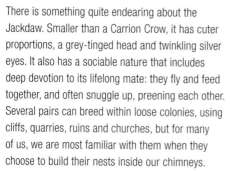

Often gives happy staccato *chak!* notes.

Some notes are longer, either *chyok* or an unhappier *charrr.*

kHz

chak   chyok   charrr

seconds

Track 48
Magpie calls

This more strikingly plumaged member of the crow tribe makes loud rattles when angry or nervous, plus slightly more musical conversational calls.

The Magpie has an air of mischief, travelling around in little gangs, flicking its long tail.

It is fairly vocal, with two calls that are loud and attention grabbing. The long rattle is the most obvious, a volley of between 5 and 20 loud *schak* calls at about 8 to 10 per second: *schak-ak-ak-ak-ak-ak-ak-ak*. It often repeats the rattle with short pauses in between, and with different lengths of rattle each time. It is a percussive noise, with no sense of pitch, given in alarm but also in irritation, and it does sound like a bird that is mightily peeved.

It could be confused with the Mistle Thrush's football rattle alarm (page 89), but the notes within the thrush's rattle are a much faster purr.

The short call of the Magpie is a more conversational and much repeated *ker-chok!* and *ker-chik!*, used to keep in touch with its mate or members of its flock. It has more overtones than the long rattle and hence sounds a little more musical; it is often given in prolonged chatter at its roost.

For the little-heard, subdued song of the Magpie, see page 202.

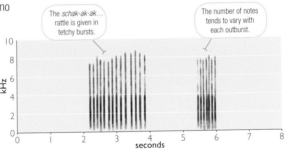

The *schak-ak-ak...* rattle is given in tetchy bursts.

The number of notes tends to vary with each outburst.

Track 49: So, to recap…

In and around even very urban areas, listen for all these species:

- House Sparrows give a primitive *chilp* and Starlings give *clicks*, *beeps* and a *weeeeeee*;
- House Martins up above let out a *preet* and Swifts scream;
- remember that Pied Wagtails have their 'Chiswick flyover' while Grey Wagtails repeat the same note *zit-zit*;
- the two crow species have their angry ˹arr ˹arr ˹arr, the Jackdaw has its *chyak!*, *chyok!* and *chyar!* and the Magpie its *schak-ak-ak*.

Track 50
Song Thrush song

Much declined, the Song Thrush is now a scarce bird in many places, but its distinctive serenades still ring out in many villages and leafy suburbs too. It is one of the stars of the show.

Brown above and with spotted underparts, the Song Thrush can easily be mistaken for the Mistle Thrush, but it is slightly smaller, with warm brown underwings. Its outer tail feathers aren't white and the spots on its breast are tear-shaped rather than rounded. Once a common garden bird, the wane in

its fortunes is thought to have been caused in part by the loss of hedgerows and wet ditches in the countryside.

The song, usually given from a perch high in a tree, is rich, bold and confident – strident even – but it is the structure that is key. The male sings a short phrase and immediately repeats it two to five times. He then pauses briefly before choosing a completely different phrase and repeats that, and so on. Maybe after three or four repeating phrases, he will insert a twiddly, longer phrase that he doesn't repeat, to throw you off the scent. This can be much more like the Blackbird. But, as with all bird songs, give him time to show his true hand. Soon enough he will return to his 'sing, repeat and move on' theme.

So a performance might run to the pattern of, 'Can you see me? Can you see me?   up here, up here, up here   high, high, high, high, high   I'm very high indeed in the tree' and so on. He can continue this for many minutes without moving from his perch.

For the rather subtle and often quiet calls of the Song Thrush, see page 222.

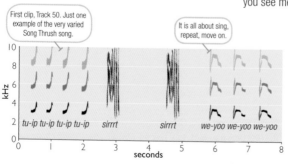

First clip, Track 50. Just one example of the very varied Song Thrush song.

It is all about sing, repeat, move on.

tu-ip tu-ip tu-ip tu-ip   sirrrt   sirrrt   we-yoo we-yoo we-yoo

So similar in looks to the Song Thrush, in song the Mistle Thrush is a lovelorn poet who then incongruously switches to a football rattle for his call.

Brown above and boldly spotted beneath, the Mistle Thrush is slightly larger than the Song Thrush, with paler markings on its face and wings. When it flies, it reveals white rather than rusty underwings, and white outer tail feathers. It likes feeding on open parkland and short turf, often hopping about far from cover but always returning to the safety of high trees.

It is here, on a lofty perch, that the male sings from midwinter onwards, the sounds carrying far. Known as the 'Storm Cock', he will sing even if bad weather is looming. The song is quite different from that of the Song Thrush, instead more closely mirroring the Blackbird's short, ever-changing, fluty verses.

However, compared with the Blackbird's verse, each Mistle Thrush verse is shorter, little more than a second long, barely three to six notes. It is deeper, simpler, with much less variation in pitch or complexity, and usually without a terminal twiddle. However, beware distant Blackbird song, where the flutier notes carry to you but the twiddles get lost.

Each Mistle Thrush verse is slightly different, so the overall effect is of mere subtle variations on a theme. It is like singing the rhyming lines of a wistful poem, such as: 'You want to go?   I let you go   So go, go now go…', etc.

The flight call is very different, like an old football rattle, *prrrrrrrrrrrt*. When agitated, the long rattle breaks into a series of short volleys, up to 10 a second, *trrt trrt trrt trrt*.

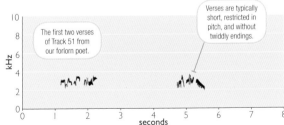

The first two verses of Track 51 from our forlorn poet.

Verses are typically short, restricted in pitch, and without twiddly endings.

Track 53: So, to recap…

- **Song Thrushes** 'sing, repeat and move on';
- **Mistle Thrushes** are lovelorn poets, reciting short lines of rhyming poetry.

Track 54
Chiffchaff song

Spring starts when the first Chiffchaff arrives and sings its name in woods and copses everywhere, making it far easier to identify by sound than by sight.

Visually, the Chiffchaff is one of those 'little brown jobs', or more accurately 'little and olive-green', a tit-sized warbler with a short, pale stripe above the eye but with few other standout features. To confidently observe the shorter wings, dark legs and other subtle features that distinguish it from the Willow Warbler requires a good view and plenty of experience.

It is predominantly a summer visitor, one of our earliest to return, with some arriving by the first week of March. However, there are also small but increasing numbers now wintering successfully in southern England. With the first burst of spring sunshine, the males break into song.

The Chiffchaff's song is perhaps one of the easiest to learn because it sounds acceptably close to its name – *chiff chaff chiff chaff*. These are simple clipped notes, slightly different in pitch, to a regular, if slightly stilted, metronomic tick-tock rhythm. The pacing is measured at about three notes per second, and this can be maintained in a verse that lasts 20 seconds or more.

However, listen closely and you will hear that the song is actually made up of three or four subtly distinct notes in a rather random order, such as *chiff cheff chaff chaff chaff cheff*. Listen, too, for males that introduce each verse with a soft, little hiccupping *h'ric h'ric*.

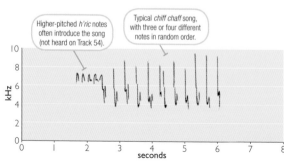

Higher-pitched *h'ric* notes often introduce the song (not heard on Track 54).

Typical *chiff chaff* song, with three or four different notes in random order.

Track 55
Willow Warbler song

While the Willow Warbler might be the spitting image of the Chiffchaff, its song could barely be more different, the lilting cascade bringing calm to spring days in many places.

Around 5 million Willow Warblers are thought to arrive here each spring from Africa. They choose open, sunny woodlands and young plantations everywhere, including in more upland areas, and especially their margins with scattered birch and willow trees. Here, they flit about in the outer branches, sometimes fly-catching and rarely still. You might be able to piece together the subtle combination of visual features to tell them from the Chiffchaff, such as the fleshy leg colour, longer wings and rather brighter and cleaner plumage tones, but it is hard.

However, the sweet song of the male is an instant giveaway. It is exquisitely soft and lilting, with no hint of harshness. This is a sound to calm your spirit. Each phrase lasts about 3 seconds and is a gently paced cascade of about 15–20 whistled notes, dropping unevenly down the scale, with a few notes often repeated a couple of times on the way down.

It might be confused with the song of the Chaffinch, which also drops in pitch this way, but for Willow Warbler think 'fairy ballerina pirouetting down the stairs' rather than a 'brash jig with ta-dah!'. Part way through the verse, the 'ballerina' sometimes jumps back up a stair or two, but the overall momentum is downwards. A male will give about four to six verses a minute, each varying slightly in composition but all very much to the same theme.

A few may sing in gardens as they pass through, but this is really a warbler of wilder places where, in many areas, it is the defining bird sound for a few weeks from April to June.

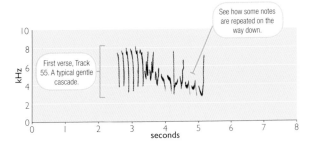

See how some notes are repeated on the way down.

First verse, Track 55. A typical gentle cascade.

Track 56
Blackcap song

This staple sound of spring woodland is one of those songs that few people know, for understanding what defines the Blackcap's 'warbles' can seem like a fine art. The secret is all to do with how this bird 'gains confidence'.

Almost ubiquitous these days in woodland, this is a rather grey warbler with a contrasting skullcap – black in the male, ginger-brown in the female. However, the Blackcap does spend much of its time in the trees, lost among the leaves, so you will hear many more than you see.

The male's main song is usually made up of warbling verses about 3–6 seconds long. The key is how the verse changes; each starts reserved and uncertain, a weak jumble of high and low notes, but it then breaks into louder, clean, beautiful fluty notes, much more stable in pitch, losing all the earlier scratchiness. So, although each verse is different, it is as if, after a shy start, the male suddenly finds confidence every time. Although the change is subtle, with practice you will increasingly perceive how most verses blossom in this way. The pace is quite fast but not out of control, and there is usually a pause of several seconds between each verse.

What make things more tricky is that the male will sometimes indulge in what is called 'long song', a longer, rambling warble full of squeaks and scratches. You may also hear a subdued subsong early in the season. A little later we will look at the Garden Warbler's song, whose similarity to the Blackcap's 'long song' continues to test even experienced birdwatchers. But for now, focus on the fact that if a warbling woodland bird 'finds its voice' halfway through the song, it is a Blackcap.

First verse, Track 56. A typical song verse, starting vague and messy.

It then breaks out into a clear, bold, simple sweet melody.

kHz / seconds

Track 57
Chiffchaff, Willow Warbler and Blackcap calls

The three common woodland warblers have consistent, short calls, rather subtle, but nevertheless very useful.

The main call of the **Chiffchaff** (below) is a fairly soft, gentle whistle, *hweet*. It has a clear upwards inflection, but it is a single syllable, simple and confident. It is often repeated but in a rather irregular manner.

The call of the **Willow Warbler** (above) is very similar, with an upwards inflection, but is distinctly two syllables – *hoo-weet*. It feels sadder, almost pitiful or apologetic.

There are a number of other species that have rather similar calls. We have already seen one of the versions of the male Chaffinch's rain call (page 76), which is very similar but bolder, more confident, and repeated ad infinitum. Later, we will also hear the Redstart's call (page 110). But for now, at least, it is a good starting point to listen for the difference between a one-syllable *hweet* and a two-syllable *hoo-weet*.

In early autumn, juvenile Chiffchaffs give a quite different call, *see-up* or *swee-oo*, dropping in pitch.

The **Blackcap**'s call is quite different, a clean, hard *tek* or *tak*, either given individually or in a more regular series. It is quite easy to mimic by clicking your tongue on the top of your palette. The sound is a little bit 'sticky' and has the character of two marbles being knocked together sharply. It is slightly but consistently different from the higher-pitched *tip* calls of the Robin and the harder flint-strike *tjik tjik* of the Wren; with practice, it is possible to tell all three species apart with confidence. All three can run the calls into a fast series: the Wren being fastest, the Robin always with that irregular dripping quality, and the Blackcap deepest and hardest.

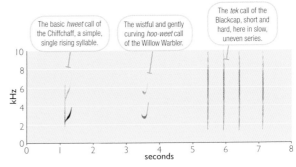

The basic *hweet* call of the Chiffchaff, a simple, single rising syllable.

The wistful and gently curving *hoo-weet* call of the Willow Warbler.

The *tek* call of the Blackcap, short and hard, here in slow, uneven series.

Track 58
Garden Warbler song

It is very easy to confuse the song of the Garden Warbler with that of the Blackcap, but there are certain subtle features to listen for.

Of all the warblers covered in this section, the Garden Warbler can be the most awkward to identify by sight, being one of those birds where its lack of standout markings is its most noticeable feature. It is basically plain brown above, buffy below, with grey legs and a stubby bill. Learning its vivacious song, and being able to separate it from the Blackcap's, is therefore all the more useful.

Like most Blackcaps, the Garden Warbler is a summer visitor and, once on territory, males sing often. Its main song is an energetic, 3–8 second warble, flowing like a babbling brook with a jumble of complex scratchy and mellow sounds. Critically, it is the 'same song' throughout, in contrast to the Blackcap's typical 'song that blossoms', and the Garden Warbler especially lacks the protracted run of pure flutiness that often ends its cousin's verses.

However, you do need to take into account the 'long song' of the Blackcap, which can be incredibly similar, and to add extra doubt some Blackcaps will sing from the kind of scrubby bushes that the Garden Warbler favours. Listen closely for a deep mellow pitch to some of the notes in Garden Warbler song, which are often Blackbird-like, fruity, dropping often to a contralto 2kHz. Also, many of its notes have a quick ripple or burr, and it often only leaves short gaps of a couple of seconds between each verse. But accept that, on occasion, these are indeed a difficult pair to tell apart with confidence.

See how it returns to rather deep, little rippling notes.

First verse, Track 58. The 'babbling brook' is too fast for our brains to keep up.

kHz — 10, 8, 6, 4, 2, 0

seconds — 0, 1, 2, 3, 4, 5, 6, 7, 8

Rightly famed for its song and for its habit of singing by night, the key identifying features of this virtuoso are power, vocal control and sheer audacity.

Everyone has heard of the Nightingale, but few have actually heard it. Indeed, it has declined by 90 per cent since the 1960s, and fewer than 5,500 singing males remain, most in south-east England and East Anglia. A summer visitor, it arrives from early April but is shy and retiring. Once a bird of coppice woodland, it is now more often found in scrubby areas near water.

Males are very vocal on their return, singing by day, when they will move regularly from perch to perch, and then it is unmated males that are vocal at night, often sticking to one perch for long periods. Most are quiet by June.

Don't expect a song of wistful melodies. Each verse is different, but is typically no more than 2–3 seconds long, most with a similar basic three-part structure: a short, twiddly introduction, then a couple of series of rapid, throbbing notes,

and finishing with an acrobatic note or two. Those series stand out: they could be a rapid-fire rattle, seesawing couplets or the sweetest whistles, all delivered with gusto, often deep and rich, and some like sci-fi laser gunfire. Some indeed get incredibly extended over many seconds.

And while variety and improvisation are hallmarks, many males include a verse that starts with a slow *tyooo tyooo tyooo* that accelerates and increases in volume over 10 seconds or more, before exploding into a rapid ending.

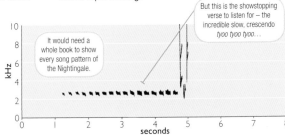

It would need a whole book to show every song pattern of the Nightingale.

But this is the showstopping verse to listen for – the incredible slow, crescendo *tyoo tyoo tyooo*...

Track 60: So, to recap…

- **Chiffchaff** simply sings its name;
- **Willow Warbler** is a graceful ballerina, pirouetting down the stairs;
- **Blackcap** and **Garden Warbler** are tricky, but if 'faltering start turns to flutiness' it is Blackcap, and if it is 'a babbling brook from start to finish' it is almost always Garden Warbler;
- the scarce, southern **Nightingale** is all about blasts of throbbing power and vocal control.

Track 61

Great Spotted Woodpecker drum and call

This is by far the most common woodpecker, which hammers on tree trunks as well as giving diagnostic calls with a kick.

A woodpecker seen at winter garden bird feeders is almost always this species. Starling-sized, it is boldly pied, with characteristic large white shoulder patches and bright red feathering under the tail. Males have a small red patch on the back of the head, which the females lack, while young birds have an all-red crown.

In spring, most head into woodland, where males and females select favourite hollow tree branches, which they visit regularly to hammer out a loud, short volley of rapid strikes called drumming. The noise carries far, each drumming-post sounding different according to the resonant qualities of the wood. There will often be a gap of 20–40 seconds before the next drum.

Each drum typically lasts for less than a second, and totals 10–20 strikes. Crucially, the drum perceptibly accelerates and becomes quieter towards the end rather than being constant all the way, like the rhythm of a flexible ruler being twanged off the side of a desk. This is subtly but noticeably different from its much rarer sparrow-sized cousin, the Lesser Spotted Woodpecker, whose longer drum maintains its strength throughout (page 198).

Throughout the year, Great Spotted Woodpeckers frequently give a very consistent and characteristic call: *kik!* or *tchik!* It is very short and sharp; indeed it is a call with a 'kick' to it. When agitated, this gives way to a fast or even very fast series of harsh *chet-chet-chet-chet...* or *chak-chak-chak-chak...* calls, loud, panicky and often falling in pitch and slightly slowing at the end.

For the calls of the begging nestlings of the Great Spotted Woodpecker, see page 198.

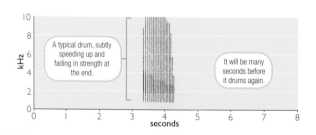

A typical drum, subtly speeding up and fading in strength at the end.

It will be many seconds before it drums again.

kHz

seconds

Track 62
Nuthatch song and call

This especially noisy woodland bird has a wide repertoire of rather liquid, piping notes that grab your attention.

There is no other British bird that can shin up and down tree trunks and branches with such consummate ease as this shuttle-shaped climber. It is tied almost wholly to areas with plenty of mature, deciduous trees, but that includes leafy suburbs and parks, and it will sometimes visit garden feeders. This makes it easier to see its attractive plumage: bluish above, strongly washed with orange below and with a thick dark line through the eye.

The Nuthatch tends to be very vocal throughout the year, with a loud, confident and varied repertoire. The quip call is the one heard most often throughout the year – a bold, piping *k'wip!*, liquid and pleasant rather than shrill. It tends to be repeated in twos, or sometimes in a rather uneven series, which can be at a rate of less than one a second, increasing to seven per second when excited. In alarm, this becomes a higher-pitched, more urgent *twip twip*....

This sets the scene for its song, of which there are two principal variants, but with many versions in between. The first is a fast, piping whinny – *wi-wi-wi-wi-wi-wi* or *quip-ip-ip-ip-ip*..., at 9–15 notes per second, and lasting 1–2.5 seconds. It is straight in pitch, well controlled from start to finish, and our bird call that sounds most like a car alarm.

The other is a slow series of whistles at a rate of two per second, which can be *tyū tyū tyū*..., or at times *weee weee weee*... or even *wee-oo wee-oo wee-oo*....

For other Nuthatch calls, see page 219. Also compare with Willow Tit song (page 208), Wood Warbler pyū song (page 213) and Lesser Spotted Woodpecker call (page 198).

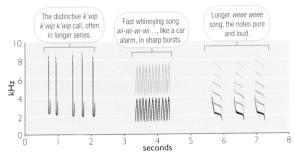

The distinctive *k'wip k'wip k'wip* call, often in longer series.

Fast whinnying song *wi-wi-wi-wi*..., like a car alarm, in sharp bursts.

Longer *weee weee* song, the notes pure and loud.

Track 63
Green Woodpecker calls

The Green Woodpecker's old country name, 'Yaffle', describes the best laugh in the British bird world, a loud, rich, drawn-out cackle. It also has a similar but shorter call.

It is always a joy to see and hear this large, red-capped, moss-green woodpecker, an ant-eating specialist and hence often seen on the ground, where it hops about perkily. However, if nervous, the Green Woodpecker will retreat to the trees with a characteristic bounding flight – the bright olive rump showing well – where it will cling in much more typical woodpecker fashion, shuffling behind the trunk to hide.

In spring, both sexes give the same advertising call, often sitting near a prospective nest hole. The 'yaffle' is a loud, attention-grabbing peal of 12–20 laughing notes, varying in the speed of delivery between six to eight calls per second, but consistent in the ringing, almost twanging timbre: *kyū-kyū-kyū-...* or *kwik-kwik-kwik....*

Each note contains complex harmonics on many pitches, and if you listen hard you can make out high, shrill overtones within the sound. Indeed, sometimes the individual notes 'crack', as if each

breaks into a higher pitch. The series is fairly constant in pitch and volume, falling slightly and sometimes speeding up a little. A bird will wait 30 seconds or more before giving the next 'yaffle'.

Throughout the year, both sexes will also give a loud call, often in flight, when flushed or on landing. This is typically two or three loud *kyak! kyak!* notes, similar in timbre to the advertising call but sharper, more hard-hitting, each note the same pitch and conveying a sense of having being rudely startled. The call can become full throttle and full volume if the bird is agitated or scared.

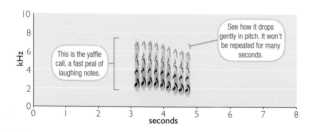

This is the yaffle call, a fast peal of laughing notes.

See how it drops gently in pitch. It won't be repeated for many seconds.

Track 64
Jay calls

This attractive bird, salmon-pink with a jaunty black moustache, broad, rounded wings and white rump, can often be quite secretive. This is where the distinctively loud, harsh call is a useful announcement that one is near, even if it stays unseen.

The Jay is predominantly a woodland bird, especially of lowland oakwoods, although it is increasingly found in some parks and gardens, where it can become quite tame. In the countryside, it remains deeply nervous of people, following a long history of persecution, although in late autumn they become much more conspicuous everywhere, as they flap languidly across the countryside in search of oak trees. Their mission is to collect acorns, carrying up to a dozen at a time in their crops; they then bury them in fields and lawns, returning to try and find them later in the winter.

The French call it *Geai*, we call it Jay, both names taken from the loud, startling, harsh screech, *SCARCH!* It is about 0.5 seconds long, so this is a long single sound compared with most bird calls. It is usually uttered singly or a couple at a time but not in a regular series. Most often it is a straight and simple noise with no music in it at all, just a blast of ultimate harshness. However, on occasion it can have a little upwards inflection and dense layers of overtones, so that you can sense a vague musical trumpet hidden among the noise.

In spring, the Jay will sometimes make a long mewing note, *mewwwwwww*, falling in pitch. This is an expert imitation of the Buzzard's call (page 100).

For the little-heard, half-hearted song of the Jay, see page 203.

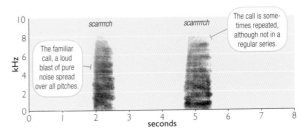

Track 65
Buzzard call

One of our most vocal birds of prey, the Buzzard has a mewing call that is one of the evocative sounds of the countryside, often made while soaring in circles on its broad wings.

Severe persecution and pesticide poisoning in the nineteenth and twentieth centuries pushed the Buzzard back to the relative safety of the hillier west and north of Britain. It was only with enlightened attitudes, stricter laws and reduced gamekeeping that it could regain lost territory, and by the early 2000s it had become the nation's most numerous bird of prey. Now it is widely seen on roadside posts or soaring the thermals, brown above, paler below, with heavy barring on the underwings and chest.

It calls in all seasons, but is especially vocal on sunny days from February to June and again in autumn. The repertoire is limited, and the call you are most likely to hear is a long *myewwwwww*. This is indeed a 'mewing' sound, more reminiscent of a cat than a human whistle. It is almost always drawn out, with a really long, sliding fall in pitch, usually almost a second long. It seems to echo around the upland valleys, somehow evoking wildness and freedom.

It is given mainly when high-soaring, especially when pairs spiral together or when neighbouring pairs meet. It can also be given from trees, although beware the Jay's excellent imitation. In aggressive encounters or when nervous, the call becomes faster, more intense and shrill, and can waver towards the end.

For comparison with the Red Kite, see page 162.

Typical long mew call, dropping a long way in pitch.

It can waver a little on the way down.

kHz

seconds

The 'sneeze call' is a way of locating a Marsh Tit and is vital to help separate it from its rare cousin, the Willow Tit, which looks almost identical. The Marsh Tit also has an angrier *bee-bee-bee* call.

Confusingly more a bird of woodlands than wetlands, a pair of Marsh Tits tends to stay on their territory much of the year, sometimes linking up with mixed tit flocks for a short while as they pass through but then returning to their more lonesome existence. They are most often to be found close to the woodland floor rather than high in the canopy. Given that the small, pale mark at the base of the bill is one of the most reliable visual features for distinguishing the Marsh Tit from the Willow, it is no wonder that the differences in their calls are so crucial for identification.

The Marsh Tit calls regularly, and has a varied vocabulary, the most frequent and useful being the sneeze call, a sharply dropping *pit-choo* or longer *pitchi-choo* or *tsi tsi choo*. It is usually given singly, with a gap before the next. Beware the ever-creative Great Tit, which can often make rather similar noises but doesn't sound quite so sneezy.

The Marsh Tit complements this with its bee-bee-bee call. It is a short series of buzzing, rather angry-sounding deep notes – *bee-bee-bee* or *zee-zee-zee* – with a distinctly nasal quality. It is often

introduced with one or two high notes, such as *chikka-bee-bee-bee*. Each 'bee' note is typically rather short, about 0.25 seconds long, so the overall call is relatively fast.

For the Marsh Tit's song, which is heard only occasionally in spring, see page 208. For how to distinguish the Marsh Tit's bee-bee-bee *call from the Willow Tit's, see page 208.*

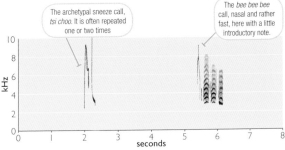

The archetypal sneeze call, *tsi choo*. It is often repeated one or two times

The *bee bee bee* call, nasal and rather fast, here with a little introductory note.

Track 67
Tawny Owl song and call

The famous hoot of the male Tawny Owl is heard only after dark, but even from urban parks. The contact call is much less known, although it is given just as often.

This is by far our most widespread owl, resident and found wherever there are plenty of mature trees. However, it is very rarely seen for it is exclusively nocturnal, while by day it is difficult to spot, so good is its incredible camouflage when roosting high in trees or indeed in tree holes. Those occasional views are often limited to glimpses of a broad-winged, robust owl in the headlights of your car, but sometimes the mobbing calls of small birds will reveal a Tawny Owl in its daytime haunt. Some are indeed tawny-coloured, others more grey, and all have a rounded face with a rather wise and knowing expression!

From autumn through until spring, a male stakes out his territory with his well-known, deep (1kHz) melancholy advertising hoot, which can be mimicked by blowing into cupped hands. It really gets going when almost all light has gone from the sky. The call has three elements: a simple, long hoot, then a long gap of 2–5 seconds before a short, quiet note, then a short pause before an even longer quivering hoot: *hooooooo* (long pause) *huh* (pause) *hūhūhūhūhooooooooo*. The calls of different males vary enough for us to be able to tell one owl from another. Females also sometimes hoot to the same overall pattern, but the notes are *much* screechier, hoarser and less controlled.

The contact call, heard throughout the year but again only at night, is a sharp *ke-wik!*, given by both males and females. Notice how it rises sharply in pitch.

For its Snipe-drum call, and the calls of Tawny Owl chicks, see page 194.

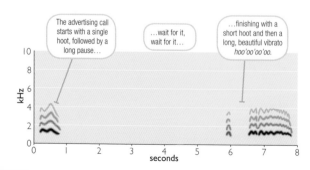

The advertising call starts with a single hoot, followed by a long pause…

…wait for it, wait for it…

…finishing with a short hoot and then a long, beautiful vibrato *hoo'oo'oo'oo.*

The endless conversation of the Long-tailed Tit's three main contact calls may be quiet, but it helps us to find a wandering troop of this most endearing of little birds.

Only distantly related to the tits, this 'flying teaspoon' has a sociable lifestyle, with extended families wandering together around their territory for much of the year and – like Pied Pipers – picking up an entourage of other tits, crests and warblers along the way.

The Long-tailed Tit is a widespread bird, mainly found in woodlands in the breeding season, using hedges, reedbeds and other tall vegetation to move from one patch of trees to the next. Increasing numbers are visiting gardens in winter.

Staying in contact is clearly vital for Long-tailed Tits, for they are persistently if discreetly vocal, using three main calls to keep the flock together, often creating a sense of conversation.

The most distinctive of its sounds is the *sirrut* ripple call, which the flock uses when anxious, when about to traverse open spaces or having spotted you! It is strongly downslurred, starting very high and dropping sharply with rapid rolling 'r's; it is also relatively loud compared with the other calls, and in high anxiety can be longer. When a troop breaks into this call, it does give the impression that they are worried.

The second sound is a high-pitched *si-si-si* call, at about 7kHz, and hence very similar to Goldcrest, Treecreeper and other tit calls. It is typically given in threes. In full alarm, it can be wound up into a faster, stronger and louder *si-si-si-si-si-si-...*, about 12 notes per second, which is much more distinctive.

The third call is the most conversational, a comfort call, short and quiet, *tuc* or *tip* or *sit*.

For the rarely heard song, see page 212.

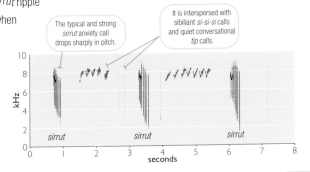

The typical and strong *sirrut* anxiety call drops sharply in pitch.

It is interspersed with sibilant *si-si-si* calls and quiet conversational *tip* calls.

sirrut   *sirrut*   *sirrut*

kHz

seconds

Track 69
Treecreeper song and calls

This well-named and unobtrusive bird frequently makes high-pitched but insistent and rather long, sibilant *seeee* notes. It also has an easily overlooked trilling song.

The Treecreeper is like an ever-climbing mouse, working its way up from the base of a trunk until it reaches the upper branches, probing at the bark as it goes, and then darting down to the base of the next tree to start again. Its brown, speckled plumage gives it excellent camouflage, and it often travels as an inconspicuous hanger-on with tit flocks.

It is a rather sedentary resident, found mainly in larger woods but also in parkland and farmland where trees and copses are well connected by hedges, for it hates to cross open spaces.

It calls regularly, but the sounds are easy to dismiss as one of the crests or tits. Most typical is a sibilant *seeee* call, usually at around 7.5kHz, which is rather loud and insistent for a call of such high pitch. Compared with Goldcrest calls, it tends to be longer, lasting about 0.25 seconds, and the sibilance is

actually a very high vibrato, which, with experience, can be picked out. Compared with Long-tailed Tit calls, the Treecreeper's calls tend to be given singly or in a slow series, maybe a couple a second, rather than in quick bursts of threes.

The song is easily 'overlooked', high-pitched and like a cross between Blue Tit and Goldcrest songs. It is a 3-second little verse, which starts with a few slow notes at a Goldcrest-high 7kHz but then breaks into a fast descending trill, ending with a small flourish that can rise in pitch, such as *shee see s'see si-hi-hi-hi-hi siup*. In contrast, the Blue Tit's trill is straight and without the flourish, while the Goldcrest's verses have a cyclical pattern.

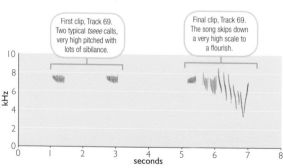

First clip, Track 69. Two typical *tseee* calls, very high pitched with lots of sibilance.

Final clip, Track 69. The song skips down a very high scale to a flourish.

This attractive and fairly rotund finch tends to be rather shy, staying deep within cover, where its soft calls are easy to overlook but offer an effective means of locating it.

The Bullfinch is usually only seen in pairs or small groups. The male in particular is stunning, with a black cap, grey back and immaculate soft-pink underparts. In the female, the pink is replaced with warm, muted grey, and at all ages the white rump is often prominent as birds flit between bushes. They stick closely to woodland with a dense understorey, and especially enjoy working their way along scrub, mature hedgerows and a few larger gardens, in spring often moving into orchards to eat fruit blossom.

Numbers rather plummeted towards the end of the twentieth century, thought to be due to changes in woodland management and the farmed environment, but they have now stabilised, albeit at a lower level.

The sound you most often hear is its contact call, which is a distinctive, if understated, sound as devoted pairs softly keep in contact with each other. It sounds like a simple $^{ph}\bar{u}$, pitiful, half-hearted and muffled. It has a vagueness to it; it seems to fade in at the start and out at the end without any real attack. It is often given in muted conversation between a pair, with one calling and then the other. Each note typically lasts about 0.2 seconds and is rather deep in pitch at around 3kHz, with a slight downslur.

For the rarely heard squeaky bicycle song, and the 'just about to fly' quiet call, see page 235.

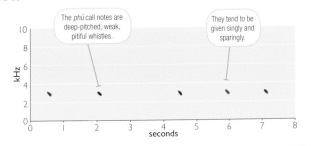

The *phū* call notes are deep-pitched, weak, pitiful whistles.

They tend to be given singly and sparingly.

Track 71
Stock Dove song

Many people could easily overlook the Stock Dove as 'just another town pigeon' that for some reason is living out in the countryside, and indeed the song of the male is rather similar.

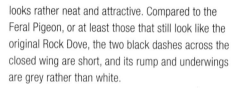

This sedentary and rather shy resident nests either in ancient rotting trees or sometimes in old barns. It tends to feed on the ground in arable fields on seeds and grain, occasionally forming large flocks. It is smaller and more compact than the Woodpigeon, lacking white neck markings and wing flashes, but it has bright pink legs and is more deeply flushed with pink and blue below, and so looks rather neat and attractive. Compared to the Feral Pigeon, or at least those that still look like the original Rock Dove, the two black dashes across the closed wing are short, and its rump and underwings are grey rather than white.

The male sings his advertising call from inside or near his nest hole, often well hidden from view. In direct display to a female, he bows repeatedly, but without pirouetting like a Feral Pigeon. It is one of those curious sounds, in that it can be heard rather faintly at some distance and yet close-to it barely sounds any louder.

The verse is a series of 4–12 deep-pitched triple notes, *whoo-er-wup*. See how it dips in pitch in the middle. A verse starts about one *whoo-er-wup* a second and then gradually accelerates, such that the final calls might be twice a second and a more disyllabic *whoo-wup*.

It would be quite easy to confuse it with the *ooooo-uh* nest call of the Feral Pigeon, so the dip in the middle of each of the Stock Dove's notes is important, as well as the habitat differences between the two species.

Spot the dip two-thirds of the way through each *whoo-er-wup*.

Each *whoo-er-wup* gets progressively slightly louder and it all gradually speeds up.

kHz

10
8
6
4
2
0

0   1   2   3   4   5   6   7   8
seconds

Track 72
Spotted Flycatcher calls and song

This supreme, dart-and-snatch insect-catcher reveals itself with subtle calls and a paltry song, but they are enough to help find one on its high perches.

The Spotted Flycatcher is a summer visitor from West Africa, still widespread but its numbers have fallen sharply. It is one of our latest migrants to return, most not arriving until mid-May, and is gone again by the end of September.

It looks like a fairly large, pale warbler, slender and usually sitting bolt upright. With binoculars, it is possible to see the large, dark eye, but it is not 'spotted' in the way the name might suggest – 'vaguely streaked' might be more appropriate. It chooses prominent positions on the sheltered outer edge of mature trees, such as dead snags, on which to sit attentively watching for flying insects. It then launches itself out on a dashing aerial sortie to catch its prey, returning to the perch it left or another nearby.

This bird is vocal but unobtrusively so, with two main notes that make up its perched call. It will frequently give a high *tsi* note, at about 7–8kHz, easily dismissed as one of many small woodland bird species. It is much more recognisable when the bird is more nervous and it adds a lower *chuk* call, or sometimes double or triple notes, giving *tsi chuk-chuk* or *tsi chu-chu-chuk*.

For the Spotted Flycatcher's rudimentary song, see page 224.

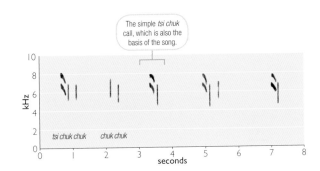

The simple *tsi chuk* call, which is also the basis of the song.

tsi chuk chuk   *chuk chuk*

kHz / seconds

Track 73: So, to recap…

The woodland is full of a mix of loud and quiet voices to add to the melodies of the warblers and thrushes. See if you can pick out the full range of species in the recording.

Track 74
Wood Warbler songs

Rarely seen away from their breeding sites, it is essential to know the Wood Warbler's two very different songs in order to find them high in the trees.

Wood Warblers arrive 'long haul' from North Africa in spring, straight onto their territories. They can be devilishly hard to find among the canopy; in appearance they are like a Willow Warbler but with a bright yellow chest and margins to the wing feathers and a pure white belly.

It means that the male's two diagnostic songs are the best way to locate them. The first, the 'spinning coin' song, is a verse lasting about 3 seconds. It is a perfectly accelerating series of *sip* notes, starting at about six notes per second and ending at 20, like the shimmering rhythm of a spinning coin as it begins to fall onto its face. The verse drops slightly in pitch as it progresses, and about two-thirds of the way through there is a subtle but definite lift in pace, as if moving to a higher gear. The male sings from favourite branches under the canopy, twisting and quivering ecstatically on his perch, or in slow flickering flights from tree to tree.

Once in a while, he will switch to his *pyū pyū pyū* song, which is a simple, gentle series of about eight or so melodic *pyū* notes. In fact, this bird's scientific name, *sibilatrix*, means 'whistler'. Each verse lasts about 3 seconds and drops slightly in pitch from start to end.

For the Wood Warbler's call, see *page 213.*

The spinning coin song gently drops in pitch, accelerating as it goes.

The *pyū pyū pyū* song is sweet, simple and relatively deep in pitch.

pyūpyūpyūpyūpyūpyūpyū

kHz

10
8
6
4
2
0

0   1   2   3   4   5   6   7   8
seconds

Found in the same woods as the Wood Warbler, the Pied Flycatcher's song is a steady little verse that uses a very limited suite of notes, which step up and down in a random order.

Arriving in April from Africa, there is no other woodland bird so smartly dressed as the male Pied Flycatcher, attired in crisp black and white as if heading out for a posh night on the town. The female is much more discreet, the black markings replaced with dull brown. In the best woods, there can be a pair every few hundred metres, often in nature reserves where scores of nestboxes have been provided for this hole-nester.

Males sing frequently on their return, using favourite branches that are often under the canopy rather than high in the crown. Each simple, bright verse uses a few basic sounds, strung together in random order one by one for about 5–10 notes to a regular beat. Each note differs somewhat in pitch and in timbre – sharp, some buzzy, some sweet.

The pacing is either 'pedestrian' or 'jogging'; it is like someone doing step exercises, going up and down three or four stairs as they fancy, occasionally repeating a couple of steps. One verse, therefore, might run *zi chū zi chū zwe zwe wer chre*, with each subsequent verse running to a different order. In terms of creativity, it is one notch up from Chiffchaff!

Once mated, males soon fall silent, so the season is short for enjoying their song (usually mid-April to late May).

For the Pied Flycatcher's wide range of simple calls, see page 225.

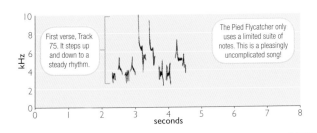

First verse, Track 75. It steps up and down to a steady rhythm.

The Pied Flycatcher only uses a limited suite of notes. This is a pleasingly uncomplicated song!

kHz / seconds

Track 76
Redstart song and calls

This dashing summer visitor to upland woodlands has an anonymous song that is difficult to learn but is well worth it, for it will help you spot many more of these beauties.

The male Redstart (above) in breeding plumage is visually stunning, with rich orangey underparts set off to perfection against grey upperparts, a black face, a pure white flash on the forehead and a rusty tail that he quivers alluringly. Females (below) and youngsters are much plainer but share the tail colour and habit; indeed, 'start' comes from the Old English word for 'tail'. A summer visitor, this bird arrives from early April onwards and its heartland is in Wales, northern England and Scotland.

The Redstart's song is often given from an unseen perch high within a tree. It usually starts with the same single note, a plaintive *soo* at about 4kHz, and then cuts straight into a confident, fast series of two to five bouncing notes. In fact, at this stage it could easily be dismissed as a Chaffinch. However, instead of then reaching the triumphant 'ta-dah!', it loses momentum, turning into a variable, short warble that runs out of steam. The ending often includes leaps of pitch and some scratchier notes, and fragments of imitation, such as Robin or Whitethroat song, but it is all rather half-hearted.

Overall, each verse is short, at 1.5–2.5 seconds, with several seconds between verses, so listen for 'the same confident start, different weak endings'.

Males and females both give two frequent calls: a plain, upslurred *hwee*, very similar to those of the Chiffchaff and the Chaffinch, and a much harder *tip* or *tik* call. It will often alternate between the two in prolonged and insistent agitation: *hwee tip hwee tip*.

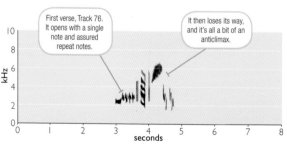

First verse, Track 76. It opens with a single note and assured repeat notes.

It then loses its way, and it's all a bit of an anticlimax.

Track 77
Tree Pipit song

If more people knew this song it would surely be a favourite of many. It is a total delight that, at its most enthusiastic and exuberant best, is one to simply stand and marvel at.

Visually, the Tree Pipit is a 'little brown job' with streaky brown upperparts and lines of bold spots on its breast. It is a summer visitor from Africa, arriving from late March. As well as being a bird of sunny glades and woodland margins in upland oakwoods, it is also a scarce breeder in native pinewoods and on heaths.

Perched high in a tree, often on top, the male sings a rapturous song. Each verse is a chain of two to four phrases, linked seamlessly, each a second or so long, so that one verse lasts 2–5 seconds. Those phrases include fast trills, seesawing couplets, whizzing repeated notes and fun, long notes that slide down the scale. So one verse might run chū-*ee* chū-*ee* chū-*ee* zi-zi-zi-zi-zi-zi zoooooo zwee zwee zwee zwee.

Often, a male will use the same opening series for every verse but vary the phrases that follow.

The male sometimes flies steeply up and then parachutes back down in song flight, adding several more series to extend the song to a joyous 12 seconds or more.

The Wren's song is also a chain of different trills and repeated notes but it doesn't have all those long notes, while Tree Pipit doesn't have the Wren's hard, rattling trills, nor does it sound so frenetic and shrill.

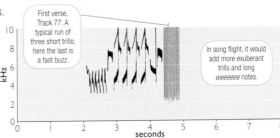

First verse, Track 77. A typical run of three short trills; here the last is a fast buzz.

In song flight, it would add more exuberant trills and long weeeeee notes.

Track 78: So, to recap…

In mainly northern and western oakwoods, listen for:

- spinning coin and *pyū-pyū-pyū* Wood Warblers;
- the 'step exercises' of Pied Flycatchers;
- 'same strong start, different weak ending' Redstarts;
- chains of joy-filled trills from Tree Pipits.

Farmland

Track 79
Rook calls

Although vocally very similar to the crows, the Rook's calls have subtle differences for you to learn.

The Rook's dense nesting colonies – rookeries – are clustered in treetops in a way that crows' nests never are. Whether here or when out feeding, it is not always easy to see the Rook's ivory-coloured skin at the base of its bill or its shaggy 'shorts' that distinguish it from the Carrion Crow, so the subtle differences in their sounds are extremely useful to know.

The basic cawing call, from which the Rook gets its name, is variable. It is typically an *arrrrr* or *corr*, straight-pitched or gently falling, in contrast to the Carrion or Hooded Crow's deeply downslurred *raarrrrk*. Some Rook calls can be a longer and downslurred *arrrrrrrr*, and some crow calls can be straight in pitch, but the extra clue is that Rooks don't tend to repeat their call in a regular series. Crucially, the heavy rasp running through the Rook's call is more pleasant than the aggressive, menacing feel of the crows' calls.

The noise of a rookery in full swing is quite an experience and it can be difficult to make out individual calls among the cacophony. However, within the basic chorus of loud *corr* notes and their variants as adult birds bow deeply at each other, it is possible to hear twice-as-high *brrup* notes, plus gurgles and rasps, which is a primitive song. As the season progresses, the squeaks and then caws of begging youngsters add to the mix.

The simple, rather pleasant *arrrr* call, often with a gentle fall in pitch amid the rasp.

Unlike the crows, the call isn't usually given in regimented repeat series.

kHz

10
8
6
4
2
0

0 1 2 3 4 5 6 7 8
seconds

Track 80
Skylark song

No other British bird hangs in the air like a drone for several minutes at a time, high above open fields, pouring its heart out in a verbal doodle without pause for breath. The 'syrupy' calls are also very useful to learn.

In looks, the Skylark is a rather anonymous pale brown bird with streaking above and below. It has a short, scruffy crest that it can erect or flatten, while in flight the wings are rather broad and triangular with a fine white trailing edge. It is a widespread resident, much declined but still with around 1.5 million pairs in the UK, with extra birds arriving from the continent for the winter. It avoids enclosed places, preferring the middle of vast cereal fields.

The Skylark sings from midwinter onwards, occasionally from a fencepost or the ground but more typically in hovering flight, rising as if on an invisible string until it is a mere speck in the sky. One song can remain unbroken for many minutes, a pleasing, burbling doodle poured out without pause for breath.

The flow can sound random, but concentrate and you will hear frequent, quick repetitions of individual notes plus short trills. Overall, despite all the deviations, there is a consistency to the timbre, as if the song has a core thread and pitch that it returns to time and again, also helped by many of the notes containing rolling 'r' sounds. It can also interweave short bursts of mimicry of other species, such as the

Swallow and the Song Thrush. The male continues to sing as he descends, falling silent a few metres from the ground.

When flushed or on migration, Skylarks also have a range of rippling, cheerful flight calls, variable but most typically a double- or triple-syllable *syrup* or *chir'r'rup*, an excellent means of locating birds passing overhead.

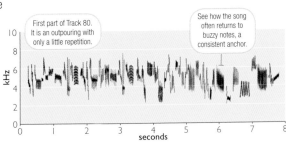

First part of Track 80. It is an outpouring with only a little repetition.

See how the song often returns to buzzy notes, a consistent anchor.

kHz

seconds

Track 81
Linnet
calls and song

Perched on farmland hedgerows, the Linnet gives an exuberant song. As with the Goldfinch and the Greenfinch, the Linnet's call is the best way to recognise the song.

Many sightings of Linnets are of a dense flock of small birds rising up from a stubble field, wheeling around and dropping back down. At a distance they look rather brown and featureless, bar a small silver flash running along their wing edges. However, a male in breeding plumage has an attractive red breast and forehead. Linnets breed semi-colonially in short, dense scrub such as hedgerows and gorse thickets, and are rarely seen in gardens and urban areas.

Flocks are very conversational, giving a chorus of their clipped *ti-dit* or *ti-di-dit* call. It is often given in flight and on take-off, and the first note is typically slightly different in pitch. The 'tickle it' calls of the Goldfinch are similar but leap up and down much more in pitch.

Linnet song is pleasing and lively, and is very varied. Some verses can be short and sparse, barely lasting 1.5 seconds, whereas 'long song' can be up to 15 seconds. The latter has a structure very similar to Goldfinch song, a series of short trills linked with quick notes. These links include low buzzes, and warped whistles that rise or fall sharply, such that a verse seems to be incredibly energetic.

This variety can make picking out a consistent theme difficult, but the *tidit* call is your golden thread, often introducing the song and then woven into and between verses. So a typical song structure is *tidit*–trill–link notes–mini-trill *tidit*–trill–link…. Resting flocks also often sing and call in chorus.

Compare the Linnet's calls with the two redpolls on page 237.

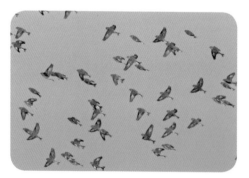

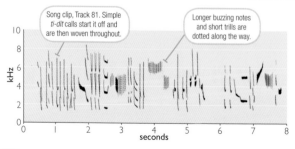

Song clip, Track 81. Simple *ti-dit* calls start it off and are then woven throughout.

Longer buzzing notes and short trills are dotted along the way.

114

Track 82
Yellowhammer song and calls

'A little bit of bread and no cheese' is one of the best-known mnemonics for any British bird's song, and effectively describes the Yellowhammer's rattle and wheeze.

Like those of many other small farmland birds, numbers of Yellowhammers have fallen significantly in recent decades, affected by the reduction in weed seeds and insects. However, the species is still widespread, if thinly scattered, through much of the UK, although it is now scarce in Northern Ireland. In breeding plumage, the male is resplendent, with a yellow head and belly and a chestnut wash on the breast and rump. Females and juveniles also have a chestnut rump but are otherwise dowdier, with hints of yellow in the face and underparts.

The male sings from hedgerows and bushes, often choosing a perch on top of the hedge or the peak of a bush within it. The initial rattle is usually about 8–12 notes at a leisurely jogging pace, and with a perceptible increase in volume and sometimes rise in pitch. The sonograms show how the notes are actually very fast couplets, rocking back and forth too quickly for the ear to separate, but it does give the rattle a sense of oscillation.

The rattle leads straight into a longer buzzing note at the end – the 'cheese' – although there are sometimes two end notes: 'little bit of bread and no cheese, please'. Each male typically has two verse versions in its repertoire.

Yellowhammers have a number of frequent contact calls, the main one to listen for being a thick, smudgy *chidd*, so formless that others have transcribed is as *stüff*, *trlp* or *tswik* – it is the 'muddiness' of the call rather than its transcription that is significant.

For other Yellowhammer calls, see page 240.

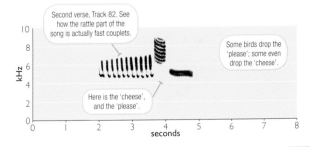

Second verse, Track 82. See how the rattle part of the song is actually fast couplets.

Some birds drop the 'please'; some even drop the 'cheese'.

Here is the 'cheese', and the 'please'.

Track 83
Corn Bunting song

This chunky bunting of arable farmland is rather anonymous in appearance, so the male's song, which is like shaking a big bunch of keys, is a standout feature.

Looking rather like a supersized, overweight female House Sparrow, the Corn Bunting is brown above, paler below and streaky all over, with a curious habit of dangling its legs in flight. Once known as the 'fat man of the barley', it has declined across much of its range, affected like so many farmland birds by the lack of seeds in winter but especially by the paucity of spring insects due to pesticides. It is nowhere abundant but can still be found in fair numbers on English downland and from Dorset up through Wiltshire and the Chilterns to Cambridgeshire, where it is resident.

Sitting on a fencepost, overhead wire or tall patch of weeds, a male sings its 'bunch of keys' song from spring right into the summer. Each verse lasts about 1.5–2 seconds. It starts with an accelerating, rising series of five to eight simple clipped *pit* or *kik* notes, as if the song is winding up. This breaks into a swirling, metallic jangle, which makes up the latter two-thirds of the verse. The jangle is uneven, often with what sounds like two or three peaks of volume, as if the bunch of keys is given a couple of swishes rather than a steady shake, and then it quickly peters out at the end.

Every Corn Bunting song sounds rather similar, but research has shown that each local population has its own subtle dialect, with all the males there singing two or three specific verses.

For the Corn Bunting's two main calls, one clipped and the other ratchety, see page 239.

First verse, Track 83. The song verse quickly winds up in speed.

The final two thirds of the verse are just a metallic jangle.

kHz

10
8
6
4
2
0

0   1   2   3   4   5   6   7   8
seconds

The Whitethroat's short 'sweet and sour' ditty is heard from hedges and bramble patches across the country in late spring and summer, sometimes expanded into a longer, sweeter song flight.

This is one of those birds that few people know, despite the male singing so frequently from his hedgerow haunts, for Whitethroats typically stay well hidden and are rather plain. Should one hop into view it indeed reveals its white throat, plus rusty wing-feather margins and white outer tail feathers. It is an abundant summer visitor from Africa, with more than 1 million pairs, although that is still less than half the number in the 1960s.

The Whitethroat usually delivers its short song from within a hedge. The song is made up of 1–1.5-second scritchy-scratchy verses, each of about 10 varied notes with a little sweetness woven in. The verses change each time and overall it can feel rather formless, and so can seem challenging to learn.

But take your time and you will begin to sense the theme within the ramble. In particular, listen how the tune lurches back and forth in pitch two or three times over the course of a verse. Compared with the Garden Warbler's song, the verses are much shorter and less rich, while the Dunnock's *diddly-diddly* doesn't have the scratchiness. There is often a couple of seconds' gap before the next verse.

In its occasional song flight, the male flutters up above its territory, singing a longer verse lasting about 3–5 seconds before parachuting back down to cover. It has the character of the 'short song' but with greater flow and sweetness, more the feel of the Skylark's song.

For the Whitethroat's zree call and alarm *churr, see page 217. Also compare it with the Dartford Warbler's song (page 217) and the Stonechat's song (page 228).*

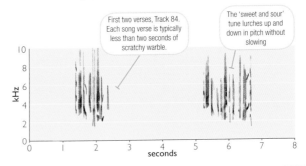

First two verses, Track 84. Each song verse is typically less than two seconds of scratchy warble.

The 'sweet and sour' tune lurches up and down in pitch without slowing

kHz

seconds

Track 85
Lesser Whitethroat song

The simple rattling song of this masked warbler is very different to the scratchy tunes of the Whitethroat, and helps find it in its high hedgerow haunts.

The Lesser Whitethroat is indeed slightly smaller than its close cousin, the Whitethroat; the main distinguishing features visually are the Lesser's lack of rufous tones in the wing and its slate-grey head and dark cheeks. It is a fairly regular summer visitor, arriving from mid-April and found mainly in lowland England but expanding its range into coastal Wales and Scotland. It prefers tall, mature hedgerows and scrub thickets.

The body of each song verse is a straightforward hard rattle of about 10–12 identical notes, lasting about 1.2 seconds, at a consistent 4kHz pitch. Listen hard and you sense that the rattle isn't just one note repeated but it actually seesaws between a main note and a very brief intervening note on a different pitch, like a highly speeded-up *chy'ok-chy'ok-chy'ok….* This is subtle, but it helps gives the song its rather mechanical character.

Perhaps the biggest risk of confusion in most parts is the Yellowhammer song, especially on those occasions where the latter fails to give its 'cheese' ending. The Yellowhammer's rattle, however, is typically slower, more relaxed, and is fuzzier or buzzier rather than the hard sound of the Lesser Whitethroat.

When you are close to a singing Lesser Whitethroat, you will also hear that it introduces each rattle with a few muttered notes, like a half-hearted warble, from which the main rattle breaks out. At the end of every verse it often leaves many seconds before the next, as it works its way along a hedge.

For the Lesser Whitethroat's call, see page 216. Also compare with Cirl Bunting song (page 240).

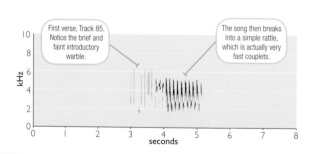

First verse, Track 85. Notice the brief and faint introductory warble.

The song then breaks into a simple rattle, which is actually very fast couplets.

Track 86
Swallow song and calls

Perched on wires near farm barns and outhouses, the male Swallow sings a sweet, fast, twittering song, including incongruous 'squeaky toy' and ratchet noises.

To think that our Swallows spend the winter somewhere in southern Africa is awe-inspiring; they may have swooped over wild elephants and giraffes before returning here! If pairs survive such epic journeys, they will return to the same nest site year after year. Seen closely, the red face and blue-glossed upperparts contrast with the white underparts, while the forked tail with long streamers is apparent.

The first Swallows arrive in late March, with most gone by early October. Unlike the Swift and the House Martin, they are rare in urban areas, much preferring farmland with plentiful livestock, among which they zip back and forth catching flying insects. In poor weather, flocks will gather to feed low over lakes and reservoirs.

The males are especially vocal in and around the nest site, singing from a perch nearby or even in flight. The song is a fast twittering of rather 'mashed' notes at a relentless, regular pace of about eight per second, such as *sitzi-tidit-chidi-tizi-ditti-chi…*. It is rather appealing and upbeat, with brief musical notes repeatedly coming through. A verse can go on for 15 seconds or can be broken into irregular chunks with short pauses.

Towards the end of longer verses it suddenly throws in a surprise, drawn-out, squeaky-toy-like *snairrr*, immediately followed by a really fast, dry ratchet *prrrrrt*, before returning to a few more twitters. A flock gathering before migration will indulge in a chorus of song.

The flight call is a brief, cheery $w^{it!}$ or w^{it}-$w^{it!}$

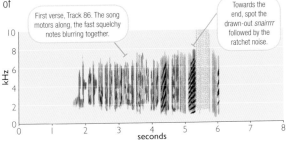

For the Swallow's alarm call, see page 211.

First verse, Track 86. The song motors along, the fast squelchy notes blurring together.

Towards the end, spot the drawn-out *snairrr* followed by the ratchet noise.

Track 87
Cuckoo song

The sublimely simple song of the male Cuckoo does exactly as the name suggests and is the essence of spring.

Rather hawk-like in shape, with sharp wings and a long tail, this summer visitor arrives from mid-April, with most adults then leaving by the end of June. The Cuckoo population has fallen by about three-quarters in the last 50 years, thought to be due to a lack of hairy caterpillars to eat and problems on their migration route.

The Cuckoo is, of course, famous for laying its eggs in other birds' nests, and so is mainly found wherever the host species live: lowland farmland for

Dunnocks, upland areas and coastal grasslands for Meadow Pipits and wetlands for Reed Warblers. It is rare to hear a Cuckoo in urban areas.

Although scarce, males do at least sing readily in spring, often from a prominent perch such as a dead tree, or even in flight. He can be heard at any time of day but especially at dusk. The ^{ku} $_{koo}$ song is easy to recognise: the two notes are distinct, the first higher in pitch than the second, with the drop the same as the ding-dong of a doorbell – musicians will know this as a 'major third'. It is repeated many times in series, with about a 1-second gap between each.

As the well-known rhyme says, 'in the middle of June' he does indeed sometimes 'change his tune', often to a rather more hurried $^{ku\text{-}ku}$ $_{koo}$ or even an off-key, rising $ku\ k'w\bar{o}$.

The only real risk of confusion is the $ku\ ^{KOOO}\ k_{\bar{u}}$ of a distant Collared Dove; reports of Cuckoos in March in the cities will be this species!

For the long bubbling call of the female Cuckoo and the choked 'gowking' of both sexes, see page 192.

The simple song is one of our best, to a simple 1-2-pause rhythm.

The notes have a soft, warm feel with the essence of summer days.

ku kū   ku kū   ku kū   ku kū   ku kū

seconds

Track 88
Meadow Pipit
song and calls

On wilder sheepwalks and moors, this singer gives what can be a monotonous and truncated song, but in parachuting song flights it lets it build into a long chain of energetic trills.

The Meadow Pipit is slim, brown and streaky, and an abundant breeder in extensive, wild and open landscapes rather than intensively farmed fields, such as downland and wet grassland but especially in summer on upland moors. Creeping around on the ground, it is rather unobtrusive although not especially shy. It looks incredibly similar to the Tree Pipit but avoids enclosed woodland and is usually loathe to perch in trees let alone sing there.

It is a vocal bird. The song starts with beautifully controlled if tedious *tsip-tsip-tsip-...* or *tew-tew-tew-...* notes, slightly sharp, at a rate of about five per second. Disappointingly, many verses are false starts, running out of energy and soon ending without having led anywhere.

However, in full song, often given in a steeply rising song flight, the opening notes gather pace as if building a head of steam until they break into a chain of trills, each of different tempo. Although the component notes of each series don't have either the whizzing quality or the long *weeeeeeee* notes of the Tree Pipit, they are still energetic and springy, some trills chinking, some rattling, and one verse can last

20 seconds or more, concluding as the bird parachutes back down to the ground, tail cocked.

The call is variations on a rather sweet, pitiful *sip* or *seep*, high-pitched (6–7kHz), in timbre slightly impure and lisping. At times, this extends into a fast series, falling sequentially in pitch, *sip-sip-sip-sip-sip-sip*.

For the Meadow Pipit's alarm call, see page 232. Also compare with the Rock Pipit's song and call (page 233).

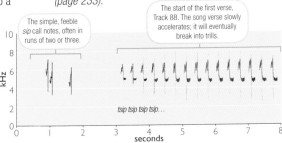

The simple, feeble *sip* call notes, often in runs of two or three.

The start of the first verse, Track 88. The song verse slowly accelerates; it will eventually break into trills.

tsip tsip tsip tsip...

Track 89
Pheasant calls

The Pheasant's very loud crowing forms a prominent sound in many parts of the countryside. Flush one and it will give a volley of panicked calls as it flies away, wings a-whirr.

Originally from Asia, the Pheasant was first brought to the UK a thousand or so years ago. Nowadays, with more than 2 million breeding females in the wild and the population boosted by the release of 35 million birds each year for shooting, this bird has become omnipresent in our countryside, strutting around field margins and roadsides. Males look striking with their chestnut plumage, red face wattles and long tail; females are duller with a shorter tail.

In an attempt to attract a harem of females, males have an incredibly short and sparsely given advertising call. It is most often heard in spring, especially very early in the morning and even before dawn. Each male finds a suitable arena on the ground, often on a slight mound, well spaced from his neighbour. He struts around slowly as if in quiet contemplation, and then suddenly explodes into a very loud, throaty, double-note crowing call of *CHARR kuk!*, neck stretched up. This is instantly followed by a rapid and loud flurry of his wings. It is several more minutes before the next blast.

All Pheasants also give a regular short call, a single sharp, loud *KOK!*, once every 2–4 seconds, or often *k'KOK!* or *k'CHOK*. When flushed or alarmed, they will clatter into the air with a rattling flurry of their wings and a rapid run of loud *KOK!* or *k'KOK!* notes, in a drama-queen volley of panic that falls away in volume as they escape and regain composure.

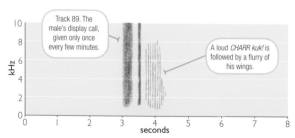

Track 89. The male's display call, given only once every few minutes.

A loud *CHARR kuk!* is followed by a flurry of his wings.

kHz

seconds

Take the rhythm of a train as it leaves the platform, cross that with a hoarse chicken, and you've grasped the chuntering advertising call of the 'Red-leg'.

This introduced resident is found throughout much of lowland Britain, although it is very rare in Ireland. About 6 million are released each year for game shooting, so small flocks (coveys) can often be seen scuttling down the margins of many arable fields. Seen well, it is easy to tell from the Grey Partridge, as it has a white face with a black line through the eye and strong 'bar coding' on its flanks, but at a distance or when running or flying away (as is often the case), identification can be much trickier.

The Red-legged Partridge can be fairly vocal in spring, with early morning and late evening the best time to hear the male's rather loud advertising call. He gives it from a low vantage point, his throat inflated. It starts slowly with low, single clucking notes, but builds up volume and speed and breaks out into a bold rhythm like the rumbling of a train down the track: *kokh   kokh   kokh kokh   k'kok-kokh k'kok-kokh   K'KOK-KOK K'CHERRR   K'KOK-KOK K'CHERRRR   K'KOK-KOK K'CHERRR*. There are variations in how the rhythm develops, but notice how the number of syllables in the repeated phrases grow.

This bird also has what is called a rallying call, a way of keeping the covey together. The sound is similar in timbre to the advertising call, but is as if the train is now into its rhythm, with a repeated *k'chook-chook k'chook-chook*. The flight call is usually given when birds are flushed, flying low and fast with an incredibly speedy whirr of wings, and often with startled hoarse calls, such as *p'chook p'chook*.

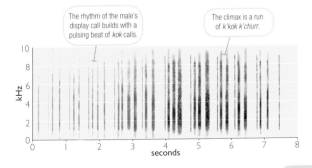

The rhythm of the male's display call builds with a pulsing beat of *kok* calls.

The climax is a run of *k'kok k'churr.*

Track 91
Grey Partridge calls

Telling the two partridges apart by sight can be difficult from a distance, so it is extremely useful to learn the distinctive calls, including the Grey's typical retching, dip-and-rise call, easily remembered as 'ear-yuk'.

Sometimes known as the English Partridge, this is our native partridge, but it is one of our most declined species. This is thought to be due to changes in the way farmland is managed which, in particular, has reduced the number of insects available for the chicks. The Grey adult is the same size and shape as the Red-legged but with a gingery face, five or six rather jagged chestnut bars on its sides and a bold, dark 'horseshoe' mark on its belly. It tends to gather in coveys of around 6–12 birds, and will run if it can to escape, or if flushed will clatter into the air, wings whirring, and fly low and fast to drop out of sight a field or two away.

It is mainly the males who give the advertising call, especially in spring, and most often at dawn and dusk or even after dark. It is a three-syllable note, about 1 second long, which is basically a scratchy retch, the first two syllables a slurred rasp that dips in pitch, followed by a clipped higher-pitched ending: *tchee-ur YUK!* or *ee-ur YUK!* You may find the words 'ear yuk' a useful reminder! The call is often given in series, one a second, and occasionally the 'yuk' ending may be weak or missing.

If flushed, the Grey often gives a rapid, panicky flight call – *hick! hick! hick! hick!...* – each note ending sharply, given in an uneven series that slows and tails off as 'danger' is left behind, sometimes finishing with double-speed 'ear-yuks'.

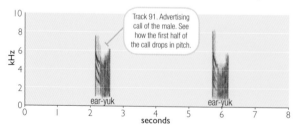

Track 91. Advertising call of the male. See how the first half of the call drops in pitch.

Track 92: So, to recap...

Listen to the track to pick out a range of common farmland birds, from rasping **Rooks** to doodling **Skylarks**, 'sweet and sour' **Whitethroats** to 'fast mash' **Swallows**. Also included are **Cuckoos**, **Yellowhammers**, **Pheasants** and **Lesser Whitethroats**.

Track 93
Reed Bunting song and calls

Here is a singer that varies the pace
of his simple song according to
whether he is mated or not.

In breeding plumage, the male Reed Bunting is dapper, his jet-black head and bib contrasting sharply with his white neck and moustache. Females and winter-plumaged males look rather more like a female House Sparrow, albeit with a more marked head pattern. Although mainly found in reedbeds and fens in summer, Reed Buntings often feed in stubble fields during the winter.

Mated males give what is called slow song. Each verse plods along, three to seven disjointed notes at a rate of two notes per second. It is like they are learning to count but getting stuck or confused. The notes sound more like German than English in timbre. So one verse might run '*ein   ein   zwei   drei*', the next '*ein   zwei   vier-vier   drei*', the last note often a short trill. Each male has a stock of 20–30 different notes it can use.

In contrast, an unmated male gives a fast song, about 8–12 notes per verse at about six notes per second, and hence he sounds a more accomplished counter! There is, however, more repetition of notes and a longer gap between each verse.

Reed Buntings also have a distinctive call. It is like saying the girl's name Sue as thinly as you can through tightly pursed lips: $^{tsi}\bar{u}$. It drops a long way in pitch, from a very high 8kHz down to about 5kHz.

For the Reed Bunting's alarm call, see page 241.

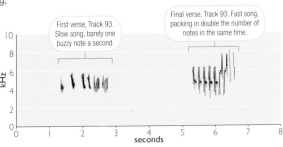

First verse, Track 93. Slow song, barely one buzzy note a second.

Final verse, Track 93. Fast song, packing in double the number of notes in the same time.

kHz

seconds

Track 94
Reed Warbler song

Separating the Reed Warbler's song from that of the Sedge Warbler is an abiding challenge for many, but there is a finger-wagging trick that will help you.

A reedbed specialist that is skilled at perching on the vertical stems, the Reed Warbler lives up to its name. It is just as happy along ditches and canals as it is in extensive stands of reed. Fleeting views are of a plain bird, warm in tone, especially on the rump, and with a white throat. It is a summer visitor, from April to September, mainly in English lowlands and south Wales.

Males sing for long periods during the breeding season, including short bursts at night. Each song verse is a series of harsh, grating churrs and sweeter notes, often unbroken for 20 seconds but up to 3 minutes long. The basic timbre is very similar to the Sedge Warbler, but it is rhythm that matters rather than the type of sound. Beating your finger in time with every note will reveal a 'Steady

Eddie', '1, 2, 3, 4…' rhythm, one note at a time, at four to five notes per second. With Sedge Warbler, your finger would be all over the place!

Yes, be aware that the beat isn't perfect; it sometimes speeds up or slows down for a few notes, and verses often start with a few seconds of excellent mimicry, often of other reedbed birds such as the Bearded Tit and Reed Bunting. However, wait and the verse will usually settle down into its comfortable rhythm. Listen, too, how the individual notes are often repeated two to five times, enhancing the pleasantly ponderous feel.

Occasionally, males will give a more conversational song, a subdued warbling version of the main song with less repetition, but this is rare.

For Reed Warbler calls, see page 214.

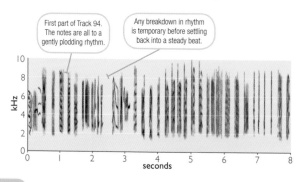

First part of Track 94. The notes are all to a gently plodding rhythm.

Any breakdown in rhythm is temporary before settling back into a steady beat.

kHz

seconds

Often confused with Reed Warbler song, the way to learn that of the Sedge Warbler is to become familiar with the complex and jazzy rhythms, not the churring notes it is composed of.

Seen well, it is easy to distinguish Sedge from Reed Warbler, especially with its bold, pale stripe above the eye, dark crown and dark-streaked upperparts. However, much of the time it remains hidden from view. It, too, is a summer visitor from Africa, arriving from early April. It is much less tied to reedbeds than its cousin, breeding in many damp habitats with dense vegetation.

The male is very vocal on arrival in spring, singing even at night. However, song ceases as soon as he secures a mate. The song is made up of a mix of harsh churring and sweeter notes in verses that can continue unbroken for a minute or more, which as a description is no different to that for the Reed Warbler's song, hence the ready confusion.

However, forget the sounds and focus again on rhythm. Sedge Warbler song is jazzy, with changes of pace and complex and jaunty patterns. During a verse, it will find a phrase it likes, repeat it, even getting stuck like a broken record, and then suddenly and radically change tack and pace. So, for example, it might go *de-de DERR, de-de DERR, de-derrr de-derrr de-derrr* and then flip straight to ^*Weeeeeeee chidididi.* You cannot wag your finger in steady time to this song! Some mimicry is interwoven, such as snatches of Swallow and Linnet calls.

A male usually sings from a slightly elevated position, in a bush or on a tall stem, but is often partially hidden. However, he also has a towering song flight, rising steeply above the territory, doing half a circle and gliding back down.

For Sedge Warbler calls, see page 214.

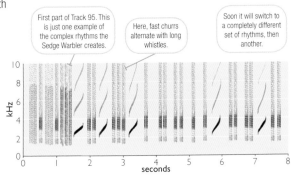

First part of Track 95. This is just one example of the complex rhythms the Sedge Warbler creates.

Here, fast churrs alternate with long whistles.

Soon it will switch to a completely different set of rhythms, then another.

Track 96
Cetti's Warbler song

The most explosive of all British bird sounds; prepare to jump out of your skin as a male belts out his sudden song from a marshside bush in mainly southern England and Wales!

Pronounced 'chettiz', this warbler is named after an eighteenth century Italian priest and zoologist. It bred for the first time in the UK in around 1972, having spread up from the Mediterranean, and its subsequent colonisation has been dramatic. Being resident and an insect eater, its population is prone to periodic setbacks if winters are harsh, but it can bounce back after a couple of good breeding seasons.

The Cetti's Warbler can be incredibly difficult to see, skulking deep in waterside vegetation, but occasionally you get a view of a rather plain, rufous-coloured warbler, with short wings and a long tail that it often cocks.

Very distinctive and easy to learn, the song is given throughout the year but most frequently and intensely throughout spring. A male dashes under cover from song perch to song perch, which are often deep within prominent willow bushes along the water's edge. At each stopping point, he gives a sudden, explosive, very loud outburst. He will often sing just once before dashing on, or may stay to repeat the verse, but only after a long pause.

Different males sing different versions, but a typical verse runs along the lines of 'chet! chet-tee! I'm a little chettee, I'm a little chettee'. You might prefer to hear it as 'chip! chip-pee! chippy-chip-shop chippy-chip-shop'. Note the single first note, then a half-second pause, then usually a couple of loud notes, a shorter pause, and then a final flourish, which often contains an element of repetition. That flourish can be quite complex or, from young birds in autumn, the whole verse can be more garbled.

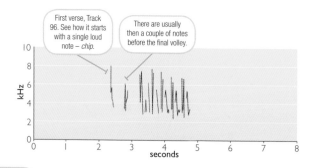

First verse, Track 96. See how it starts with a single loud note – *chip*.

There are usually then a couple of notes before the final volley.

Track 97
Grasshopper Warbler song

This mouse-like bird creeps about through the dense undergrowth of summer wetlands, only giving itself away with its incredible, grasshopper-like song.

The Grasshopper Warbler is a scarce and declining summer visitor from Africa, and a visit to its strongholds in west Wales, west Scotland and Ireland may be needed to hear it. Males can sing for long periods in spring and are at their most vocal from dusk till dawn.

The reeling song is a fast mechanical rattle, indeed like a grasshopper stridulating or a fishing reel unwinding, most unexpected for a bird. Indeed, it is so unbird-like that it would be easy to dismiss it as an insect. The sound is slightly tinny, at a constant and fast 25 notes per second. The pitch is rather high, focused at around 5.5kHz, and it carries far across the marsh. Each verse takes several seconds to reach full volume and then usually continues for at least 20 seconds without apparent pause for breath, but this is a bird that at times can sing for an hour without pause. However, at the end of a verse the sound 'switches off' instantly.

It can be difficult to pinpoint the source of the sound because the volume changes each time the singer turns its head, giving the song a ventriloquial quality. It can even be difficult to tell how close the bird is to you. However, with luck the sound may guide you to him, perched unobtrusively on a stem, beak wide open, throbbing with the effort.

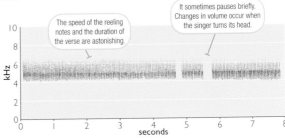

The speed of the reeling notes and the duration of the verse are astonishing.

It sometimes pauses briefly. Changes in volume occur when the singer turns its head.

Track 98: So, to recap…

In reedbeds and marshes, listen for:

- **Reed Buntings** trying hard to count in German, quickly or slowly;
- 'Steady Eddie' chuntering **Reed Warblers** and jazz drummer **Sedge Warblers**;
- 'chippy chip shop' **Cetti's Warblers**;
- and the fishing line reeling of **Grasshopper Warblers**.

PART 3
Reference Guide

The sounds of birds of the UK and north-west Europe

This part covers more than 250 of the most frequently encountered bird species seen and heard in the UK and north-west Europe, detailing the main songs and calls of each.

Remember that songs and calls can vary markedly, especially in some species. Also, individuals differ from one another, and a bird can change its vocalisations according to its mood or situation. The acoustics of a location and how far away the listener is from the source of a sound also affect how songs and calls sound (page 22). While that means this reference section can only act as a guide, the descriptions nevertheless set out the main sounds you can expect these species to make.

I have highlighted species with particularly interesting or amazing songs and calls by introducing them as 'Star Sounds'.

Brent Goose

The rolling deep *brrup* calls of flocks is the epitome of eastern and southern estuaries and marshes in winter.

CALLS Very vocal when landing and taking off. This is perhaps the most pleasant of all goose sounds, quite deep, restrained, warm and conversational without much of the show-off honking or shrill *wink*-ing of other geese. It is a noise that can be mimicked quite successfully, just by saying *brrup* with a good roll of the tongue on the 'r's and rising slightly in pitch. As usual with geese, different birds utter the call at their own pitch, such that a large flock creates a rippling and loud chorus.

COMPARE WITH Barnacle Goose.

Branta bernicla

UK STATUS Winter visitor and passage migrant, mainly October–April. The dark-bellied race of around 100,000 birds winters mainly in eastern and southern England; the pale-bellied race includes 3,400 birds in Northumberland and 36,000 in Ireland.

HABITAT Coastal saltmarsh, estuaries and grazing marshes.

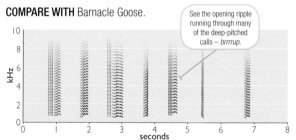

See the opening ripple running through many of the deep-pitched calls – *brrrrup*.

Canada Goose

This familiar goose is an introduction from North America but is now totally at home on town lakes and reservoirs everywhere.

***MMM-RUK!* CALL** Very vocal and very loud. This characteristic call starts with a quiet, deep, short hum, then breaks sharply into a much higher-pitched, loud bugling note: *mmm-RUK!* or *mmm-RONKH!*; the more excited they are, the louder and higher the *RUK!* In flocks, the *mmm* can be hard to make out, leaving a cacophony of different-pitched *RUK!* calls.

DUET Pairs proclaim territory with triumphant duetting, with one bird calling and the other immediately responding, slightly higher in pitch, up to four notes per second (two notes each): *ONKH ANKH ONKH ANGK*. The calls don't have such a nasal quality as Greylag Goose calls but are still easy to confuse.

Branta canadensis

UK STATUS Resident across most of lowland Britain, although some birds make semi-migratory journeys within the country in autumn. The population is more than 62,000 pairs and growing.

HABITAT Generally lowland lakes and gravel pits, including in city centre parks, feeding out on short grasslands and arable crops including amenity grasslands and golf courses.

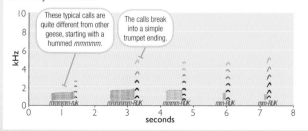

These typical calls are quite different from other geese, starting with a hummed *mmmmm*.

The calls break into a simple trumpet ending.

Barnacle Goose

Once entirely a bird of wild northern places, their yapping can now be heard from feral populations.

CALLS Very vocal in flight and especially on take-off and landing. Most often heard as a flock, in which the chorus of calls creates a cacophony, easy to confuse with that of other geese. However, you should be able to pick out a mix of low-, medium- and high-pitched calls, each of which is typically one short syllable. It doesn't have the *wink wink* type calls of the Pink-footed or *ANGKH ANGKH* of the Greylag, but instead *ruk!* and higher *rak!* calls, sounding rather like the irritating yaps of a small dog, underlain with quiet *rō!* or *rrrō!* calls.

COMPARE WITH Pink-footed Goose; bean geese; White-fronted Goose.

Branta leucopsis

UK STATUS Winter visitor, from end October–April, with 94,000 visiting a few key locations in mainly Scotland and western Ireland. There is also an increasing feral resident population with almost 1,000 pairs in lowland Britain.

HABITAT Wintering populations visit coastal marshes and estuaries; feral birds use gravel pits and wetlands.

WHERE TO HEAR Lindisfarne (Northumberland); Caerlaverock (Dumfries and Galloway); Islay (Argyll); Minsmere (Suffolk).

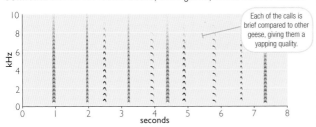

Each of the calls is brief compared to other geese, giving them a yapping quality.

Greylag Goose

A mix of wild and feral populations of this hefty orange-billed goose trumpet nasally on lakes across much of the country.

CALLS Very vocal. The calls are variable but most tend to be loud and nasal. It is the first sound on Track 10. There are some typical calls to listen for: a double *ANGKH ANGKH*, where each note can either be straight, or can break in pitch. The call may also have a shorter second note, *ANGKH agh*, or extend to *ANGKH agh agh*, or be rattled off at double speed, or as simply *agh agh agh*. It also has a very long and pained *AAHHHHHHNNNG*. Notice how the same basic underlying syllable runs throughout almost all the variations.

COMPARE WITH Canada Goose; bean geese; Pink-footed Goose.

Anser anser

UK STATUS Widespread and increasing feral population plus, in the far north, a native resident and winter visitor from Iceland. In total, just under 50,000 breeding pairs, with more than 230,000 individuals in winter.

HABITAT In the far north a bird of lochs and coastal moors; further south typically found on gravel pits and reservoirs, feeding on farmland.

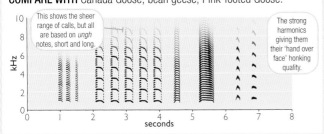

This shows the sheer range of calls, but all are based on *ungh* notes, short and long.

The strong harmonics giving them their 'hand over face' honking quality.

Taiga and Tundra Bean Goose

Previously considered as one species, it has recently been split into two, based on small differences in bill coloration, size and breeding range – the Taiga Bean Goose (right) winters on the Buckenham Marshes, Norfolk, and the Slamannan Plateau, Falkirk,

Anser fabalis
Anser serrirostris

while the Tundra Bean Goose (far right) turns up less predictably and in very small numbers. Both bean geese tend to be less vocal than other 'grey' geese, and are indistinguishable from each other by sound.

CALLS Very similar to the Pink-footed Goose, with single and double *ung* and *ank* notes, but they are typically deeper, more 'beefy', without such high-pitched, shrieky *wink wink* calls.

UK STATUS No more than 300 in total of the Taiga Bean and a few dozen Tundra Beans are seen each year.

HABITAT Agricultural fields, wet grassland and marshes.

Pink-footed Goose

Huge flocks in a few special locations create one of our winter sound spectaculars.

STAR SOUND: CALLS Very vocal in flight and especially on take-off – large flocks are often audible at considerable distance. Overall, the sounds can be difficult to tell from other grey geese, but through the cacophony, it is nevertheless possible to pick out distinct and very frequent double calls, typically a mix of medium-pitch *ungh-ungh* and very high-pitched and piercing *wink-wink* calls. It is the fourth sound on Track 11. In comparison, Greylag calls don't include the *wink-wink* higher notes, White-fronted include a squeaky bicycle *wil-a-wik* and the notes of the rare bean geese are rather deeper.

COMPARE WITH White-fronted Goose; bean geese.

Anser brachyrhynchus

UK STATUS Winter visitor, late September–April, with about 360,000 individuals.

HABITAT Mainly coastal grazing marshes and nearby arable fields, flying to roost overnight on saltmarshes, reservoirs and estuaries.

WHERE TO HEAR Norfolk coast; Martin Mere (Lancashire); Westwater Reservoir (Scottish Borders); Loch Leven (Perth and Kinross); Montrose Basin (Angus); Loch of Strathbeg (Aberdeenshire).

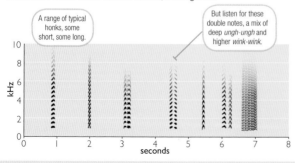

A range of typical honks, some short, some long.

But listen for these double notes, a mix of deep *ungh-ungh* and higher *wink-wink*.

kHz

seconds

White-fronted Goose

A very localised wild winter goose, whose 'squeaky wheel' calls are subtly different to those of other geese.

CALLS Very similar to other wild geese; there is a typical mix of high *wik* calls and lower honks, similar to the Pink-footed and bean geese, but in White-fronted Geese some higher notes often sound fractured, more *wil-a-wik* than *wik*; they are so high they often have a squeaky-wheel quality, the notes halfway between musical and painful! En masse, you get a sense of a riot of different noises, with a wider range of pitches than in other geese, the dominant high notes underpinned by occasional deep bass notes.

COMPARE WITH Pink-footed Goose; bean geese.

Anser albifrons

UK STATUS Winter visitor, October–March, mainly to a few traditional haunts, with about 20,000 Greenland birds but only 2,000 (and declining) of the European race.

HABITAT Greenland birds visit wild bog country and low-intensity farmland; European birds mainly use coastal grazing marshes.

WHERE TO HEAR Isle of Sheppey (Kent); Slimbridge (Gloucestershire); Buckenham Marshes (Norfolk); Holkham (Norfolk); Loch Ken (Dumfries and Galloway); Loch Gruinart (Islay).

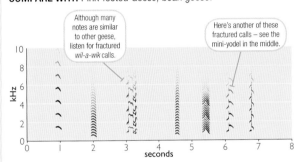

Although many notes are similar to other geese, listen for fractured *wil-a-wik* calls.

Here's another of these fractured calls – see the mini-yodel in the middle.

Mute Swan

This familiar huge white bird of town lakes is quiet but not mute, but its standout sound is that of its wingbeats.

CALLS Pairs and groups of birds make quiet, deep, conversational snorts and sneezes, heard only at close range, such as the third sound on Track 7.

THREAT CALL Threatened birds, especially with young nearby, hiss in an intimidating manner, often with threat posturing or even in lunging attack.

SINGING WINGS In flight, the wings beat a consistent three times per second making a distinctive pulsing, singing sound, full of beautiful harmonics, audible from a kilometre away and rich in Doppler effect as they pass by. Birds flying together synchronise their wingbeats, accentuating the sound.

FEET SLAPPING Its giant webbed feet slap across the water surface on take-off, or when coming in to land. It makes the most of this in territorial disputes, with dramatic splashy charges at intruders, the Mute Swan equivalent of a Gorilla beating its chest.

COMPARE WITH Bewick's Swan; Whooper Swan.

Cygnus olor

UK STATUS Resident, with more than 6,000 pairs spread across almost all lowland areas, and a winter population of almost 80,000 individuals including many not yet at breeding age.

HABITAT A familiar sight on large lakes, canals and rivers where there is ample aquatic vegetation. Also grazes in grass fields and young arable crops.

Bewick's Swan

Scarce and declining winter visitor, now found in just a few traditional locations, where its bugling calls are a defining sound of winter.

STAR SOUND: CALLS Fairly vocal, especially in flight, when the combined noise can be an excited and constant chorus, the sound carrying far. Its loud calls include short, clipped *ag ag* yapping notes and rich, harmonic, longer bugling. Some have the shrillness of a child blowing a toy trumpet; others are like the honking of a squeezy horn on a vintage car. In comparison to the Whooper Swan, the calls are on average slightly higher in pitch, and more barking and 'goose-honk' than pure trumpeting, but the differences are slight.

WING NOISE The wingbeats are audible but quiet, just a slight whistle.

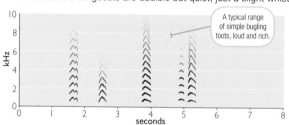

A typical range of simple bugling toots, loud and rich.

Cygnus columbianus

UK STATUS Winter visitor from Arctic Russia at a small number of sites, most from late October–late February. Probably now only 2,000–4,000 individuals.

HABITAT In winter, visits extensive flooded grasslands plus arable and stubble fields, with open water nearby to retreat to for roosting.

WHERE TO HEAR Slimbridge (Gloucestershire); Welney (Norfolk).

Whooper Swan

Named after its loud call, this is part of the soundscape of northern lochs in winter.

STAR SOUND: CALLS Heard especially in flight, on landing or from herds on the water, the main call is a loud, sonorous *WUP! WUP!* trumpet, similar to the Bewick's Swan but a little louder, deeper, more musical and less yapping. It is often given in quick doubles or trebles. Birds call at different pitches, so a herd sounds like a slightly off-key brass chorus.

TRIUMPH DUET When breeding, pairs duet as they greet or having seen off a rival. First one then the other utters a longer *WHOOP!*, becoming increasingly synchronised, as they head-bob and stretch their wings.

WING NOISE Just a quiet whoosh of air.

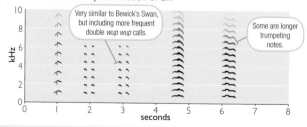

Very similar to Bewick's Swan, but including more frequent double *wup wup* calls.

Some are longer trumpeting notes.

Cygnus cygnus

UK STATUS Winter visitor with about 15,000 individuals, mainly November–March. More widespread than the Bewick's Swan, with mainly small herds visiting many northern lochs and meres. The Ouse Washes attract very large numbers. Very small numbers breed in Scotland.

HABITAT Shallow wetlands, including lowland flooded grassland and lochs, moving onto fields to feed.

Egyptian Goose

As numbers of this black-eyed introduced goose grow, there are more opportunities to hear the males' strange steam-engine calls.

STEAM-ENGINE CALLS (MALE) A regular series of identical, hissing long notes, *chrrrrt chrrrrt chrrrrt...*, as if he has lost his voice, and like sharp bursts of steam from a steam engine. It is usually at a consistent two notes per second but sometimes rattles along at 8–12 notes per second, *cht-cht-cht-cht-cht....*

METRONOMIC HONK (FEMALE) No other goose honks in such a steady, unstoppable series, *onk onk onk onk onk...*, often two per second for many seconds, sometimes breaking into a rapid-fire 10 honks per second, that sweeps up in pitch and volume and then dies away again.

Alopochen aegyptiaca

UK STATUS Sedentary introduced resident, mainly in East Anglia and London/Surrey area but now spreading, with about 1,100 pairs.

HABITAT Reservoirs, gravel pits and smaller pools, and coastal marshes. Often found in outdoor pig rearing areas and nests in early spring in large tree holes.

WHERE TO HEAR The Serpentine (London); Rutland Water (Rutland); Norfolk Broads area and north Norfolk coast.

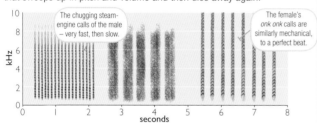

The chugging steam-engine calls of the male – very fast, then slow.

The female's *onk onk* calls are similarly mechanical, to a perfect beat.

Shelduck

This large, tricoloured duck gives curious calls, especially the males as they chase females in spring courtship flights.

CALLS (MALE) Mainly used February–May, the sound is unexpectedly feeble and high-pitched, like the cheeps of a duckling, with an airy, whizzing quality, *wheesp wheesp...*, or run into a faster series of *psup-psup-psup-psup....*

CALLS (FEMALE) A loud, clucking series, *ag-ag-ag-ag-ag-ag...*, starting slowly but quickly accelerating to a full intensity cackle, five to six notes a second, then tailing off. At times it can be even faster, or a slower, more deliberate, upslurred *rrrrUP rrrrUP...* or *grraa grraa....* Mixed with the male's calls, they create an odd combination.

Tadorna tadorna

UK STATUS Thinly scattered, mainly around the coast, with about 15,000 breeding pairs and 66,000 individuals in winter. Many adults move to the Heligoland Bight (Germany) in summer to moult their flight feathers.

HABITAT Largely coastal on muddy and sandy estuaries, but a few breed on gravel pits and dry, sandy heaths.

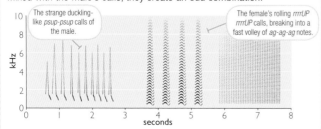

The strange duckling-like *psup-psup* calls of the male.

The female's rolling *rrrrUP rrrrUP* calls, breaking into a fast volley of *ag-ag-ag* notes.

Mandarin

This introduced duck from the Far East is our gaudiest, the male impossibly ornate in his breeding plumage, with bold white eye-stripes set in fans of golden cheek feathers, and triangular orange 'sails' on his back. It is generally a very quiet duck, little heard except at the start of the breeding season.

CALL (MALE) In early spring, with their fancy head plumes and ginger 'whiskers' inflated, groups of males display to each other and the females, flicking their heads and giving rather weak, upslurred, high-pitched whistles, *wheee*. It is accompanied by a tiny sneeze, only audible at very close range.

CALLS (FEMALE) A short, repeated bark, *brak brak*.

Aix galericulata

UK STATUS Sedentary resident, with more than 2,300 pairs, most in southern and central England but spreading.

HABITAT Extensive woodlands with shady pools, breeding in tree holes and owl boxes.

Garganey

This is a much sought-after but scarce and retiring summer visitor, the males stylishly attired with a striking white crescent either side of their heads. Their dry rasp calls may help you locate them in the margins of vegetated pools.

CALL (MALE) A distinctive hollow wooden rattle, *prrrrt*, totally dry, similar to running your thumbnail down a short length of comb.

QUACK CALL (FEMALE) May occasionally quack, like a female Mallard, only weaker, higher in pitch and more like a squeaky toy in timbre.

Spatula querquedula

UK STATUS Summer visitor and passage migrant from Africa, with a wide scatter of records in spring but fewer than 100 nesting attempts each year.

HABITAT Shallow, well-vegetated pools.

WHERE TO HEAR Hotspots include larger wet grassland sites such as the Somerset Levels and Ouse Washes (Cambridgeshire/Norfolk).

Shoveler

Never abundant, this distinctive duck with its oversized bill is easy to identify, especially the dapper males in their breeding plumage with their chestnut sides and green heads, but they have undistinguished, paltry little calls that are rarely heard.

CALL (MALE) The display calls are a double *bok-bok* call, weak and clipped, like conversational chicken noises.

QUACK CALL (FEMALE) A rather short, slightly nasal quack, not full-bodied or laughing as with the female Mallard.

Spatula clypeata

UK STATUS About 500–1,000 pairs breed, joined in winter by northern European birds with a total of about 18,000 individuals.

HABITAT Shallow, vegetation-rich, freshwater lakes and flooded grasslands in the lowlands.

Gadwall

Drake Gadwalls 'stand out' for the relative dullness of their plumage compared with most other ducks. The calls, too, are rather understated, but with practice are quite distinct and well worth learning.

'BIB AND WHISTLE' CALLS (MALE) Most vocal during courtship in late winter and spring. Very typically gives a clipped, rather apologetic *bib* call, like a half-hearted honk on a mini-horn, often repeated at a rate of about one a second. Much more easily overlooked is the weedy, short, high whistle used in courtship, often interspersed with the *bib* call.

QUACK CALL (FEMALE) Gives straight-pitched, rather restrained *wak* calls, like an embarrassed female Mallard. When it occasionally extends into a 'decrescendo' laughing run of *wak* calls, it is often only for three or so notes, compared with the 10 or so of the Mallard, and is rather raspier than the open-throated guffawing of its cousin.

COMPARE WITH Mallard (male and female); Teal (male and female).

Mareca strepera

UK STATUS Resident breeder with about 1,200 pairs, plus a winter visitor from Iceland and Europe, boosting numbers to about 25,000 individuals.

HABITAT Shallow gravel pits and other lowland waters with plenty of submerged vegetation. They often feed on plant titbits brought to the surface by diving Coots.

Teal

Our smallest duck is elusive when breeding, but the *peep*-ing males are a feature of winter marshes.

PEEP CALL (MALE) In late winter and spring, small groups of males gather to display on shallow pools, twisting and turning sharply as they swim. One male after another will nod sharply, then head toss and squeeze up their body, accompanied by a freely given, far-carrying, rather high-pitched *peep*; at close range listen for a slight ripple at the start, *preep*. The call rises slightly in pitch, giving it an expectant feel. Each is short and simple, typically at a rate of about one a second, but beware the similar calls of the Mallard and the Gadwall.

QUACK CALL (FEMALE) A weak, rather pathetic quack, like a high-pitched female Mallard, often in quick series.

COMPARE WITH Pintail (male); Gadwall (male and female); Mallard (male and female); Garganey (female).

Anas crecca

UK STATUS About 2,000 pairs breed, mainly in north Scotland. Much more widespread in winter when about 220,000 birds visit from northern Europe.

HABITAT Breeds on undisturbed, well-vegetated pools, often in the uplands. In winter, most abundant on flooded grasslands but also on estuaries, lakes and gravel pits.

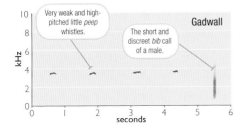

Very weak and high-pitched little *peep* whistles.

Gadwall

The short and discreet *bib* call of a male.

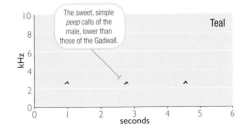

The sweet, simple *peep* calls of the male, lower than those of the Gadwall.

Teal

Wigeon

Wigeon are like avian sheep, grazing in large, tight flocks on the short turf of damp winter grasslands. It is a social and vocal duck, with the male's far-carrying, whizzing whistle given freely throughout the winter, a defining sound of many wetlands and one that will make you smile every time.

STAR SOUND: WHIZZING CALL (MALE) The high, excited, far-carrying whistle is heard by night as well as by day. It is quite pure in timbre – just a simple breathy whipped blast on a little Swanee whistle, hw^eo_o, rising from 1kHz up to 4kHz and back down again. These whizzing whistles ripple around a large flock, and are made as they woo females but also to keep in contact with their mates and when posturing to one another. They are often interspersed with little *hwip* calls, like hw^eo_o calls that don't quite have the energy to whizz upwards.

GROWLING CALL (FEMALE) This is a gruff and rather inconspicuous very deep, rumbling growl, *brrrr* or *bru-brrrr*, sometimes repeated again and again when being chased in flight by overzealous males, or uttered in duet with her mate.

COMPARE WITH Eider (males).

Mareca penelope

UK STATUS Mainly a winter visitor to lowland Britain from northern Europe and Russia, with around 450,000 birds, the first arriving in late August and some staying until April. Just 300–500 pairs stay to breed on Scottish lochs.

HABITAT In winter, favours large open wetlands such as broad river valleys, levels and washlands, plus estuaries, moving onto extensive short turf where they can graze but retreating to the safety of the water if nervous.

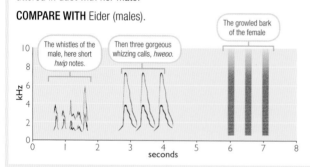

The whistles of the male, here short *hwip* notes.

Then three gorgeous whizzing calls, *hweoo*.

The growled bark of the female

Pintail

Drakes are incredibly elegant, but this is rather a quiet, shy wintering duck of wild places, and it is often too distant to be able to hear the male's faint display calls.

DISPLAY CALLS (MALE) In group display, small groups swim around performing ritualised headshakes, tail flicks and bottom lifts. As they do, they utter little whistled *prrip* calls, easy to confuse with the Teal, squeezed in between a feeble whizzing wheeze.

QUACK CALL (FEMALE) The occasional kazoo-like quack is similar to that of the female Mallard but rather short.

COMPARE WITH Teal (male); Mallard (male and female); Gadwall (female).

Anas acuta

UK STATUS Winter visitor, with just under 30,000 individuals visiting a relatively small number of traditional wintering locations. About 30 or so pairs breed in northern Scotland.

HABITAT Mostly found on sheltered estuaries and flooded wet grasslands.

Mallard

Our most familiar duck, often tame and found on ponds and pools everywhere, has a range of expressive and often loud calls. It is especially vocal January–April, and early morning and evening.

RAAB CALL (MALE) Used as a contact and alarm call, the muffled *raab* is like a muttered, husky quack. It is often slightly upslurred and has a hint of a frog croak. It becomes *ra-raab* in aggressive encounters with other males.

DISPLAY WHISTLE (MALE) During group courtship displays, males swim around each other, abruptly rearing up in the water and flicking their tails, giving a very simple, rather quiet whistle, *pyü*, easily confused with the male Teal but even higher pitched, followed by a quiet grunt. It sounds too high for a large duck.

QUACK CALL (FEMALE) A full-bodied, very loud *MWAK! MWAK! MWAK...*, often given in a series of 10 or so notes, about two a second, descending in pitch like heartily laughing down the scale, towards the end speeding up slightly and losing volume. However, it sometimes is delivered in a straight-pitched, slow series of *mwak* notes, sometimes as few as one a second, continuing without a break for 20 seconds or more.

COMPARE WITH Gadwall (male and female); Teal (male and female); Pintail (male and female).

Anas platyrhychos

UK STATUS Resident and winter visitor across almost all of Britain, with more than 100,000 breeding pairs and more than 700,000 individuals in winter.

HABITAT Almost any wetland including brooks and rivers, in the breeding season preferring those with shallow water where the ducklings can find invertebrates, and nesting on vegetated islands, in dense bankside cover or in tree holes.

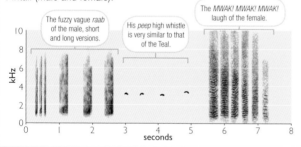

The fuzzy vague *raab* of the male, short and long versions.

His *peep* high whistle is very similar to that of the Teal.

The *MWAK! MWAK! MWAK!* laugh of the female.

Red-crested Pochard

This chunky diving duck has established a small feral breeding population in just a few parts of the country, where you have a chance of hearing their occasional sounds. Males are distinctive, with a large, velvety golden orange head and bold red bill.

CALL (MALE) The display call is very curious, like a fizzing sneeze, *quiz*, as you might imagine a duck quack would sound on helium.

QUACK (FEMALE) The flight call is a deep, dog-barking *rruk rruk rruk*.

Netta rufina

UK STATUS Small feral population with only about 20 breeding pairs and a winter population of fewer than 500.

HABITAT Largely on gravel pits.

WHERE TO HEAR The Cotswold Water Park (Gloucestershire/Wiltshire/west Oxfordshire) is the most reliable site.

Pochard

Although most obvious in winter in deep water pools, seek out their more secluded breeding sites to hear the male's fun display call.

DISPLAY CALL (MALE) Generally silent, but in display the male makes a brilliant if very quiet Swanee whistle, *whee^p* or *whee-wowww*, not unlike that of the male Eider but longer, wheezier; it whips up, or up and down, often accompanied by a deep *nug nug nug*.

CALL (FEMALE) A rolling *brrarr brrarr*, mainly given in flight, similar to the Tufted Duck but dipping in pitch at the end.

Aythya ferina

UK STATUS Mainly a wintering species with a declining population of around 48,000 birds. About 500 pairs stay to breed but are secretive.

HABITAT In winter, found on favoured reservoirs and gravel pits; breeds on densely vegetated lowland pools.

Tufted Duck

First breeding in the UK in 1849, this is now our most widespread diving duck, the black males with their white side panels especially easy to spot. However, they are vocally rather reserved.

DISPLAY CALL (MALE) Silent for most of the year, in late winter and spring small groups of males make cute bubbling giggles to woo females. The notes are high and fast as if speeded up; imagine cartoon mice sniggering. The sound is very quiet and difficult to hear at any distance.

CALL (FEMALE) A simple, short, rasping *grrrr grrrr grrrr*, straight in pitch and repeated once or twice a second in short series.

COMPARE WITH Pochard (female); Red-crested Pochard (female); Goldeneye (female).

Aythya fuligula

UK STATUS Resident breeder across lowland areas with just under 20,000 pairs, winter visitors boosting the population to about 120,000 individuals.

HABITAT A familiar sight on freshwater gravel pits, reservoirs and lakes, even sometimes on city park lakes.

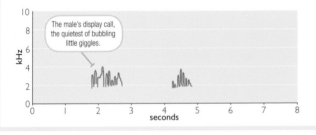

The male's display call, the quietest of bubbling little giggles.

Scaup

Much rarer than the Tufted Duck and a winter visitor only, it feeds on marine molluscs offshore before heading back to breed on mountain and tundra pools in northern Europe.

CALLS The quiet calls are rarely audible and perhaps little uttered but the male's display giggles are similar to those of the Tufted Duck.

Aythya marila

UK STATUS Scarce winter visitor with only about 12,000 birds.

HABITAT Scottish firths and Northern Irish loughs.

Eider

This hefty, sociable sea duck, with a wingspan of more than a metre, is a familiar sight in harbours and sheltered bays in the north, allowing close-up encounters with displaying males as they make the campest of bird calls.

STAR SOUND: DISPLAY CALL (MALE) One of the most amusing, indeed saucy, of all bird calls, most often heard in late winter and spring. It is a smooth, deep, simple, rise-and-fall $o^{oo}h$ note, sounding rather human. However, every now and then the thrust is bigger, the rise in pitch higher, so that it sounds like they have just spotted something rather risqué: $ahh^h{-}OOOH$. It isn't especially loud, but it carries well across the water, and flocks of males give it in a glorious, titillated chorus.

CALLS (FEMALE) Females make *uk-uk-uk* or *og-og-og* calls, sometimes loud but more conversational when reassuring their crèches of ducklings, and often continuing for several seconds.

Somateria mollissima

UK STATUS Mainly sedentary resident, with about 27,000 breeding pairs, almost all from Northumberland northwards, with a few in Northern Ireland and Walney Island (Cumbria). A few disperse to winter around coastlines further south.

HABITAT Cold, sheltered, coastal waters rich in shellfish; especially abundant around islands.

WHERE TO HEAR Sheltered bays around Northumberland and the Scottish coast, especially the Ythan Estuary (Aberdeenshire), Orkney and Shetland.

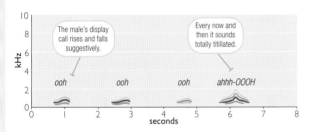

The male's display call rises and falls suggestively.

Every now and then it sounds totally titillated.

ooh *ooh* *ooh* *ahhh-OOOH*

kHz

seconds

Velvet Scoter

This dark sea duck, like the Common Scoter but with a bold flash of white at the rear of the wing, is not a vocal bird and is almost never heard in the UK. It is only on the northern European forest lakes where they breed that you might hear its rudimentary and occasional barks and whistles.

Melanitta fusca

UK STATUS About 2,500 birds winter.

HABITAT Exclusively at sea, mainly in Scottish firths.

Common Scoter

Most of the time, these sea ducks are seen as distant flocks bobbing in the winter swell or flying in fast, drilled lines low over the sea. Not especially vocal and often distant, it is difficult to hear their calls.

CALLS Males make a plaintive chorus of rather weak *pyū* whistles; females infrequently give grating barks.

Melanitta nigra

UK STATUS Rare breeder in Scotland; 100,000 winter around the coast.

HABITAT Moorland lochs in summer; shallow seas in winter.

Long-tailed Duck

This beautiful but scarce sea duck should perhaps be called the singing duck. Males, resplendent in their art deco, white-and-grey plumage and streamer tails, are a particularly attractive duck in spring, although getting close enough to hear them is a challenge.

DISPLAY CALL (MALE) Often gathering in small flocks, males in late winter and spring give joyful, rising yodels in a toy trumpet voice, both in display and in flight. Musicians will recognise the call as an arpeggio in a major key: *up u-wa up*, in distant chorus likened to bagpipes.

OTHER CALLS Both sexes also have more clipped, conversational *wup* or *wuk* notes.

Clangula hyemalis

UK STATUS Winter visitor, with around 11,000 birds, mainly in Scotland.

HABITAT Sheltered sea lochs and firths.

WHERE TO HEAR Moray Firth (Highland); Forth of Forth (Fife/Lothian); and, in spring, Loch Belmont on Unst (Shetland).

Goldeneye

This dashing, deep-diving duck, males boldly marked in black and white and both sexes with an 'inflated' triangular head shape, is unusual in that it has 'singing wings'. The male's curious display rasp is also worth listening for.

WING WHISTLE In flight, the wings of males, and to a lesser extent females and juveniles, produce strong, fast-pulsing airy whistles, 10 per second.

DISPLAY CALLS (MALE) In late winter, small groups of males swim frantically around the females and one will suddenly arch his neck in a double-jointed backbend while making a strange percussive sound: *ber-BEECH-rr*. It has a dry, rasping timbre, like a ratchet with a hint of kazoo! Males also make a quieter, deep *brrrrrrr brrr brrrrrrr…*, rather irregular in length, like winding up a clock.

CALL (FEMALE) Has a hard barking *grra grra* or *brrr brrr*, sometimes in rapid sequence.

COMPARE WITH Tufted Duck (female); Pochard (female); Red-breasted Merganser (female).

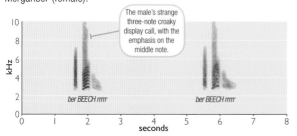

The male's strange three-note croaky display call, with the emphasis on the middle note.

ber BEECH rrrr   *ber BEECH rrrr*

Bucephala clangula

UK STATUS Winter visitor, with a population of only about 27,000 in Great Britain. About 200 pairs stay to breed, mainly on Speyside and using special nestboxes.

HABITAT In winter, visits larger reservoirs and lochs, plus coastal waters including some estuaries. Breeds in woodland, fishing on deeper stretches of rivers and lochs.

WHERE TO HEAR In the breeding season: the River Spey and surrounding lochs (Highland). In winter: Abberton Res (Essex); Rutland Water (Rutland); Forth Estuary (Fife/Lothian); Loch Leven (Perth and Kinross); Loughs Neagh and Beg (Northern Ireland).

Smew

This little diving duck is an exciting visitor from northern Europe. The stunning white-and-black adult males are outnumbered by the number of 'red heads' (females and first-winter males) that reach us in winter. However, their calls are almost never heard here, and even on their breeding grounds they apparently only make a few dry rattling, croaking and creaking sounds.

Mergellus albellus

UK STATUS Scarce winter visitor with barely 100 each year, mainly in south-east England.

HABITAT Tends to turn up on gravel pits where there is deep water to dive in.

Goosander

This fish-eating athlete of a diving duck has expanded its range and is now a much more familiar sight than previously. Males are strikingly creamy-white with a black head and back; females are very similar to Mergansers. Like its cousin, this is not a vocal duck and is rarely heard.

DISPLAY CALLS (MALE) Small groups gather together, tossing their heads and scooting across the water, giving rudimentary creaks and double *burg-barg* notes, the first note higher than the second, or a faster *b'bog*.

CALLS (FEMALE) A deep, Raven-like *brōk* or fast *buk buk buk buk* series.

Mergus merganser

UK STATUS About 3,500 pairs breed in mainly Wales, northern England and Scotland, with 12,000 individuals in winter more widely spread.

HABITAT Breeds along upland rivers in summer, spreading more widely to preferred lowland lakes and reservoirs in winter.

Red-breasted Merganser

This slender but rather large and long-necked diving duck, with its just-out-of-bed spiky crest, is normally seen rather distantly and its calls are infrequent and quiet, making this one of the more difficult British birds to hear.

DISPLAY CALLS (MALE) During the amazing group display, males swim together, periodically stretching their necks forwards and half submerging in dramatic 'curtseys', before 'power paddling' across the water. As they do this, they make quiet *buk* notes followed by pathetic wheezing sounds, like a high-pitched Mr Punch. However, you will need to be close to hear them.

CALL (FEMALE) Makes a simple *brrrowk*, similar to the female Tufted Duck but deeper and more guttural.

Mergus serrator

UK STATUS Fewer than 2,000 pairs breed, mainly in the north and west, with a midwinter population of about 9,000 more widely spread around the coast and including some northern European birds.

HABITAT Estuaries, coastal lochs and inshore waters.

Capercaillie

The display call of this huge woodland grouse is extraordinary but, given its perilous UK status, you should only seek them on guided events.

DISPLAY CALL (MALE) Males gather at dawn in early spring to display, first in the trees and then down on the ground. Slow, hollow, paired *tok* sounds accelerate, culminating in a 'cork pop', followed by a strange noise, like someone scribbling furiously with a scratchy pen on paper.

BELLOW CALL (MALE) Especially during the evening, males give a belching *eurrgghhh!*, like a small bellowing deer.

CALL (FEMALE) A deep muffled quacking, *kok! kok! kok!*

WING NOISE On take-off, the wings make an immense wind-rush.

Tetrao urogallus

UK STATUS Sedentary and much declined resident, restricted to small areas of the Scottish Highlands, with only 1,100 individuals in 2015/16.

HABITAT Caledonian pine forests and plantations.

WHERE TO HEAR Check if Loch Garten is running Capercaillie dawn watches; otherwise, please respect the calls to steer clear of the forests in the early morning to let them lek in peace.

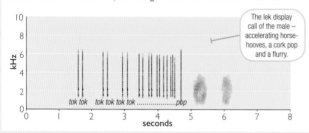

The lek display call of the male – accelerating horse-hooves, a cork pop and a flurry.

Black Grouse

Looking like portly, black velvet chickens, males gather at dawn in spring to strut and call at their communal display – the lek.

STAR SOUND: LEK DISPLAY CALL (MALE) Their beautiful, rhythmic bubbling calls carry far, each cycle of the call lasting about 3 seconds, rising in pitch, and running to the rhythm of *bubalub bu'blub bu'blub*. However, in chorus and at a distance it sounds like a throbbing bubble. Each bird periodically breaks into a very different, throat-clearing hiss, *chew-ishhh*. It is like there is a cat among the pigeons!

OTHER CALLS Females make various gentle *bok bok* clucking sounds plus a rapid *cuk-cuk-'uk'uk'uk'urr* that falls in pitch.

Lyrurus tetrix

UK STATUS Sedentary resident, much declined, with probably fewer than 5,000 lekking males remaining in undisturbed parts of northern England, Scotland and Wales.

HABITAT Moorland edge and bogs in remote areas, often with scrubby young woodland and plentiful Bilberry.

WHERE TO HEAR Highly prone to disturbance; there are very few places you can go to listen to Black Grouse, so join a guided trip, mainly in Highland Scotland.

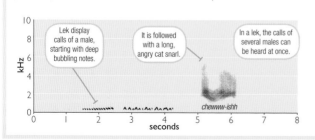

Lek display calls of a male, starting with deep bubbling notes.

It is followed with a long, angry cat snarl.

In a lek, the calls of several males can be heard at once.

Ptarmigan

This attractive, high-mountain grouse has three main calls, all rasping, like running a stick along notches in a hollow tube – its name comes from the Scottish Gaelic *tàrmachan*, meaning 'croaker'.

FLUSHED CALL (MALE) *Arrrrrk ark r-arrrrk*, to the rhythm of 'Here comes the bride'.

FLIGHT DISPLAY CALL (MALE) The male takes off, rises steeply, glides briefly, and when descending calls *ARRRK ak-arrr*, followed by *rk-rk-rk*.

PERCHED CALL (MALE) This extended rattle dips in pitch as it starts, slows a little, and then rises and accelerates into a long, loud ending: *row-ARRRRK*.

Lagopus muta

UK STATUS Sedentary resident in the Scottish Highlands with around 8,500 pairs, although numbers vary from year to year.

HABITAT Found mostly on the highest mountains, preferring areas with very short vegetation, rocks and scree.

WHERE TO HEAR The Ptarmigan Restaurant on Cairn Gorm (Highland), or else a long slog up a remote Highland mountain.

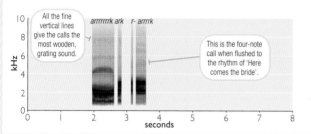

Red Grouse

A walk on the moors is all the more entertaining on hearing the males' two different advertising calls.

STAR SOUND: TOWERING DISPLAY CALL Especially at dawn and dusk in spring, males make short flights on the edge of their territories. They rise steeply with a loud *BOWK!* and then parachute slowly down, making a decelerating clucking noise. On landing, they call *g'back! g'back!* or *g'bowa g'bowa*. This is the first call on Track 13.

PERCHED ADVERTISING CALL Starts with slow clucks and then winds itself up, *boork bork bak bak bak bak bukukuk'k'k'k'k'k'k*, rising to a peak in volume and pitch, and then dropping away at the end.

Lagopus lagopus

UK STATUS Resident, with about 230,000 pairs, mainly in northern England and Scotland, with small numbers in Wales, south-west England and Northern Ireland.

HABITAT Heather-clad moors and mountains, preferring open areas without trees. Large areas of moor are managed in some parts for Red Grouse shooting.

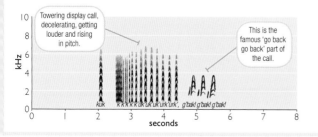

Red-legged Partridge

Now by far the commoner of the two partridge species, the 'steam train' call of this non-native gamebird is often heard on farmland.

'STEAM TRAIN' ADVERTISING CALL See page 123 for full description. A male stands on a favourite low vantage point and starts slowly with low, single, well-spaced clucking notes, *kokh   kokh   kokh*, but builds up volume and speed, the notes becoming double *k-kokhs*, before breaking out into a loud, excited crowing, *K'KOK-KOK K'CHERRRR K'KOK-KOK K'CHERRR*, all to the rhythm of an accelerating train rumbling down the track.

RALLYING CALL A repeated *k'chook-chook k'chook-chook*.

FLIGHT CALL When birds are flushed, they fly low and fast, often making startled hoarse calls such as *p'chook p'chook*.

ALARM CALL A repeated, grating, flat-pitched call, *chhhhht*, like a steam train pulling up an incline.

COMPARE WITH Pheasant (call); Egyptian Goose (male).

Alectoris rufa

UK STATUS Non-native, introduced resident throughout much of lowland Britain (although very rare in Ireland), with more than 80,000 territories, plus 6 million are released each year for game shooting, temporarily boosting the population.

HABITAT A farmland bird, mainly on arable but also pasture and moorland edge.

Quail

If you are lucky, this elusive, tiny gamebird reveals its presence in summer from dense cereal fields with its repeated 'wet-my-lips' call.

STAR SOUND: ADVERTISING CALL (MALE) Most often heard June–August and mainly at dawn and dusk, the call is three notes to a set rhythm: *wip w'wip* (pause) *wip w'wip* (pause). It is easy to see how this has become known as 'wet my lips'. He gives it stretching upwards, his head sharply jolted with each note. He will usually repeat this verse half a dozen times or so, and then break for several minutes. While the noise isn't strong, it carries far. The whole song is introduced with a quiet sore-throated *wah-wih* noise, usually given twice, as he inflates his throat.

FLUSHED CALL Much less familiar, both sexes sometimes make a *wrreeee* call as they take flight.

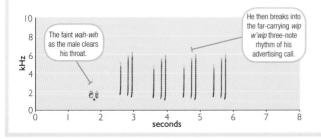

Coturnix coturnix

UK STATUS Scarce summer visitor, wintering in Africa. Numbers vary each year with maybe 500 singing males in a good year, with the largest influx in 2011 when there were around 2,000.

HABITAT Large, open fields with cereal crops, enjoying the warm southern chalk landscapes but often venturing as far as Scotland.

WHERE TO HEAR Although there is no guaranteed location, the Dorset, Wiltshire and Berkshire Downs are perhaps the best areas.

Grey Partridge

Track 91

Telling the Grey Partridge from the Red-legged Partridge by sight can be difficult from a distance, so learning the Grey's distinctive 'ear-yuk' call is useful.

'EAR-YUK' CALL See page 124 for full description. The advertising call is a scratchy two-syllable retch. It starts with a slurred rasp that drops in pitch, followed by a clipped higher-pitched ending: *tchee-ur YUK!* or *ee-ur YUK!*

FLUSHED CALL If flushed, it gives a rapid *hick! hick! hick! hick!…* in uneven series that slows and tails off, sometimes with double-speed 'ear-yuks' on landing.

COMPARE WITH Red-legged Partridge.

Perdix perdix

UK STATUS Sedentary and widespread resident in lowlands, now scarce in most areas, with around 43,000 territories. In places, it is still released for shooting.

HABITAT Mainly arable farmland, especially with weedy stubble fields, field margins and 'bird cover' crops.

WHERE TO HEAR Arable farmland often close to the coast, especially in the West Sussex downs, north Norfolk, Lincolnshire and Yorkshire.

Pheasant

Track 89

Originally from Asia, this long-tailed gamebird is a prominent voice of farmland and woods. See page 122 for full description.

DISPLAY CALL In spring, males pace around a suitable arena and every few minutes suddenly explode with a very loud, throaty crowing *CHARR kuk!*, followed by a rapid flurry of their wings.

SHORT CALL A single, sharp, loud *KOK!*, once every 2–4 seconds, or often *k'KOK!* or *k'CHOK!*.

ALARM CALL When flushed, birds clatter into the air giving a rapid run of loud *KOK!* or *k'KOK!* notes, easing as they reach safety.

COMPARE WITH Red-legged Partridge.

Phasianus colchicus

UK STATUS Introduced resident, with more than 2 million breeding females, but the population is boosted by the release of 35 million birds each year for shooting.

HABITAT Farmland, especially where copses and bird cover crops are managed specifically for them.

Golden Pheasant

Introduced from the mountains of China, the male is one of the world's most extravagantly plumaged birds, mainly bright red beneath and gold above but with feathers of every other colour, and a Pharaoh's headdress. However, they are shy and retiring, so the calls may be all you encounter.

ADVERTISING CALL (MALE) A quick two-note wretch, *arch-etch!*, the second note a little higher pitched than the first.

Chrysolophus pictus

UK STATUS Introduced gamebird, with fewer than 100 pairs remaining in places such as the Brecks (Norfolk).

HABITAT Typically dense, dark thickets in woods, venturing out at dusk.

Red-throated Diver

The eerie, drawn-out duets of our smallest diver echo over the water in their wild haunts and are said to signal impending rain.

WAIL CALL During March–June, the powerful drawn-out wails rise and fall in a gentle arc: *owwwwwwwwww*. Often a pair's calls overlap in a tormented duet.

STAR SOUND: 'ARROW-UP' CALL At the height of excitement, the female breaks into a powerful, *ah-RRROO-up! ah-RRROO-up!*..., the middle note vibrato. The male will duet discordantly with her to a faster rhythm, his call deeper, with a growling first syllable that breaks into a much higher-pitched second: *rrrr-wup! rrrr-wup*.

OTHER CALLS May give short, simple barks in alarm and short quacks, *wark wark wark*, in flight.

Gavia stellata

UK STATUS About 1,300 breeding pairs, joined in winter by migrants from northern Europe, boosting numbers to around 17,000 birds.

HABITAT Breeds on remote, undisturbed lochs in northern Scotland, within easy flight of the sea for feeding. Winters around the British coast, generally within sight of shore.

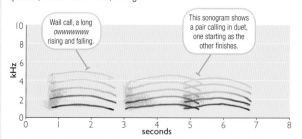

Wail call, a long *owwwwwwww* rising and falling.

This sonogram shows a pair calling in duet, one starting as the other finishes.

Black-throated Diver

This rare bird is prone to disturbance, so keep your distance should you encounter a pair on the northwestern Scottish lochs where they breed.

STAR SOUND: *WUP-WOOOO-EE* CALL The main territorial call is a short series of steeply rising wails that break in pitch like a yodel and are linked by a quick *wup* note: *woooo-ee   wup-woooo-ee   wup-woooo-ee*.

WAIL CALL Also used territorially, each note rises slowly then has a whipped or yodelled ending: *wooooooooo-oh!*

FLIGHT CALL A deep, resonant frog-croak, either short or long.

Gavia arctica

UK STATUS Rare breeder in north-west Scotland, with only 220 pairs. In winter about 500 individuals are spread thinly around the coast, with hotspots off western Scotland and south-west England.

HABITAT For breeding, uses large freshwater lochs with plenty of fish and nesting islands. In winter, visits large sheltered coastal bays.

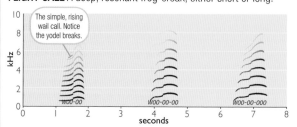

The simple, rising wail call. Notice the yodel breaks.

woo-oo   *woo-oo-oo*   *woo-oo-ooo*

Great Northern Diver

This bulky, impressive diver only rarely makes its amazing, loud and far-carrying calls in the UK.

CALLS The wail call is similar to that of the Black-throated Diver, but is sometimes followed by a rhythmic, yodelling *Wee-oo-wah, Wee-oo-wah*. It also has a vibrato laugh, wild, wobbly and very beautiful.

Gavia immer

UK STATUS Winter visitor to north and west coasts, with around 2,600 birds.
HABITAT Mainly offshore with odd singles turning up on large reservoirs.

Storm Petrel

This tiny seabird, the size of a House Martin and similarly black with a white rump but with a black belly, only comes to its remote island nests after dark, when its strange electric purr-and-gulp calls can be heard from deep between boulders and stone walls. For the incredible experience of hearing them call, you need to visit one of their few breeding sites after dark, the most accessible being the dusk boat trip to the island of Mousa (Shetland), where Storm Petrels breed in the ancient broch (stone tower).

STAR SOUND: PURR-AND-GULP OCCUPATION CALL Heard at the nesting site after dark from April–August, both sexes 'sing', with the impression that there are endearing little elves within their rocky lairs. It is a perpetual purr with a machine-like, almost electronic feel, every 1.5 seconds interrupted by a quick whimpered gulp as if grabbing air, before the purr starts again, creating an unending cycle: *prrrrrrrrrrrr-chow-prrrrrrrrrrrr-chow-....*

DISPLAY CALL Either at the nest or chasing each other in the dark, birds repeat a squeaky *k-cherr tik!*

Hydrobates pelagicus

UK STATUS Only comes ashore to breed in a few locations on the western seaboard of mainly Scotland and Ireland, with about 25,000 pairs. It winters in the eastern Atlantic, as far south as South Africa.

HABITAT Even in the breeding season, it stays beyond sight of land by day, feeding on plankton, only returning after dark to its remote islands, where it nests in rock crevices or in dry stone walls.

Leach's Petrel

Slightly larger than the Storm Petrel, it too is a bird of the open ocean, only seen from shore when a few windblown individuals pass by after intense autumn storms. Few people get to hear its cute calls at its remote breeding sites.

CALLS There are two main calls. The first is a buzzing trill, *brrrrrr*, that rises in pitch, then breaks into a higher-pitched, rising *weeee*, followed by a little hiccup, *brup*, which is all then repeated in a cycle: *brrrrrrrrrrr-weeee-brup-....* The second is a smile-inducing, jaunty ditty; imagine a fast sing-song, 'I'm a little Leach's, and I'm in my hole' rhythm, given in a speeded-up Woody Woodpecker cartoon voice.

Oceanodroma leucorhoa

UK STATUS The breeding population is thought to be about 50,000 pairs in fewer than 10 colonies, the largest on St Kilda (Outer Hebrides).

HABITAT Scree at the foot of cliffs of some of the most inaccessible islands.

Fulmar

These stiff-winged, grey-and-white seafarers are a noisy presence at their cliff-ledge nests but are otherwise rarely heard.

CALLS Nesting pairs, which can be on territory from early winter, sit face to face and 'sing' to each other, bills wide open, throats inflated, heads nodding, or shout at rival birds flying too close. The calls are a variety of loud, throaty clucks and cackles, from deeper, slow raspy moans to rapid *uk* notes, usually in a long series that grows in intensity and then settles down but can quickly escalate again.

COMPARE WITH Gannet (call).

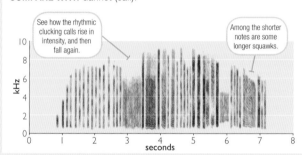

See how the rhythmic clucking calls rise in intensity, and then fall again.

Among the shorter notes are some longer squawks.

Fulmarus glacialis

UK STATUS About 500,000 pairs breed, the majority in northern Scotland with smaller numbers further south. In winter, they disperse locally out to sea.

HABITAT Spends most of its time at sea, only coming to land to breed on high sea-cliff ledges or on gentle clifftop slopes on predator-free islands.

Manx Shearwater

Most sightings of these wanderers are of loose lines of birds, far out at sea, barely beating their stiff, long wings as they tilt one way then the other. You need to make a special trip to one of their few island colonies to hear the incredible chorus of nocturnal 'wheezy chicken' calls.

STAR SOUND: CALLS After dark in spring and summer, males and females return to their breeding burrows, creating an unforgettable mass of sound. The calls are an excited and asthmatic crowing, in three- or four-note repeated rhythmic cycles such as *k'kok-kok-herrr, k'kok-kok herrr*, the last note sounding like a deep intake of raspy breath. Each individual's call has its own unique rhythm.

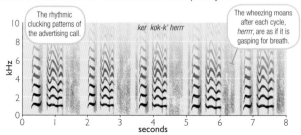

The rhythmic clucking patterns of the advertising call.

ker kok-k' herrr

The wheezing moans after each cycle, *herrrr*, are as if it is gasping for breath.

Puffinus puffinus

UK STATUS Breeding visitor to about 50 colonies, totalling 300,000 pairs, about 80 per cent of the world population. They winter in the southern Atlantic off South America.

HABITAT Spends almost its entire life at sea, out of sight of land. Breeding colonies are on remote, predator-free islands, where they nest in earth burrows.

WHERE TO HEAR Skomer and Skokholm islands (Pembrokeshire); Rum (Inner Hebrides).

Little Grebe

This diminutive grebe, like a floating powder puff, often hides from view among pond vegetation but gives itself away with its cheeky calls.

ADVERTISING TITTER This fun-filled sound is a rapid titter, up to 6 seconds long – it is the third sound on Track 12. It rises in pitch and volume near the start, before gently dropping in pitch and intensity in a long 'tail' as if running out of steam. It is an infectious and rather naughty giggle. It is often given in duet by a pair, and can be accompanied by frenetic paddling across the water as they chase off rivals.

ALARM CALL A series of very clipped, high *wik!* notes

COMPARE WITH Slavonian Grebe.

Tachybaptus ruficollis

UK STATUS Widespread resident, with more than 5,000 breeding pairs across lowland areas.

HABITAT In the breeding season, uses freshwater pools, often small and always fringed with extensive vegetation such as reeds. Many disperse in winter, some moving onto coastal wetlands or rivers.

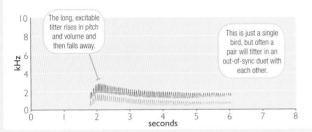

The long, excitable titter rises in pitch and volume and then falls away.

This is just a single bird, but often a pair will titter in an out-of-sync duet with each other.

Great Crested Grebe

This elegant water bird is better known for its elaborate courtship than its calls, but its occasional barks and growls can be an unexpected spring sound on lakes and reservoirs.

CALLS Vocally rather reserved but it actually has a wide repertoire of calls, some short and staccato, some long and moaning. All tend to be rather deep and can be nasal, barked, rattled or trumpeted. Most tend to be given in a short series before falling silent again. In particular, it gives a vibrato growl half a second long, which can be deep or a trumpet, often dropping slightly at the end, *grarrrrrrr*. Later in the season, chicks peep persistently to beg for food.

Podiceps cristatus

UK STATUS Resident, with just over 5,000 breeding pairs across lowland areas. In winter, arrivals from Europe boost the winter population to about 23,000 individuals.

HABITAT Breeds on reservoirs, lakes and gravel pits with a good margin of vegetation. In winter, mainly found on larger reservoirs with some around the coast.

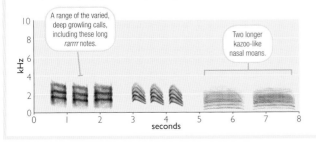

A range of the varied, deep growling calls, including these long *rarrrr* notes.

Two longer kazoo-like nasal moans.

Red-necked Grebe

Numbers visiting our coasts have slumped, with probably fewer than 50 birds in total each year. Given that it is largely silent in winter, and summer records are very scarce indeed, there is little chance of hearing its cackling display calls.

Podiceps grisegena

UK STATUS Scarce winter visitor.

HABITAT Mainly in sheltered, shallow coastal waters.

Slavonian Grebe

Sadly, this stunning small grebe with golden 'horns' is declining in its last Scottish breeding haunts, the only place to hear its calls including its tittering duet. In winter it is largely silent.

ADVERTISING/ALARM CALL A slurred call with a simple slight rise and then falling arc, *ee^errrgh*, distinctive in that only the second half of the call is full of rolling 'r's.

DISPLAY CALL A fast, bubbling, variable titter that typically rises quickly in pitch and volume, and is often given in pulses rather than the long, unbroken titter of the Little Grebe, winding down into a protracted slower 'exhausted' end, such as *titterteer titterteer titteer titteer tee, teer*. Usually a pair will titter in duet.

COMPARE WITH Little Grebe.

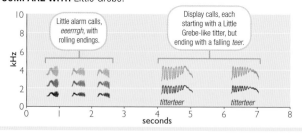

Little alarm calls, *eeerrrgh*, with rolling endings.

Display calls, each starting with a Little Grebe-like titter, but ending with a falling *teer*.

kHz

titterteer   titterteer

seconds

Podiceps auritus

UK STATUS Very rare summer visitor, declining and now fewer than 30 pairs; winter visitor from Iceland and Norway with around 1,100 individuals.

HABITAT Breeds on sheltered lochs with emergent vegetation; winters on shallow, sheltered coastal waters, typically in a few favoured locations such as Pagham Harbour (West Sussex).

WHERE TO HEAR Loch Ruthven (Highland).

Black-necked Grebe

This rare grebe, in breeding plumage with a stylish spray of metallic golden feathers behind its eye, is rarely heard.

CALLS Only vocal in the breeding season, and even then it is difficult to hear the calls over the cacophony of the Black-headed Gull colonies it tends to nest among. Its main call is a simple two-syllable *zwee-tuc!*, usually repeated several times, rather like the Stonechat. It also gives an occasional trilling display call in short pulses, weaker than the Little Grebe.

COMPARE WITH Stonechat; Little Grebe.

Podiceps nigricollis

UK STATUS Fewer than 50 pairs breed at about 20 widely scattered sites in England. The main wintering sites are along the English south coast.

HABITAT Breeds on lakes with plenty of floating vegetation. In winter, visits sheltered estuaries and bays.

Spoonbill

Unmistakable and well named, this all-white heron-like bird has black legs and a ridiculously long, spatulate bill. It has long been a scarce, non-breeding visitor to the UK but is now starting to colonise, with pairs breeding in Norfolk and Yorkshire, and increasing numbers staying for the winter with Poole Harbour the main location. However, it is almost entirely silent throughout the year, even at its breeding sites.

CALLS The occasional calls are merely low-volume grunts and mutters that are very difficult to hear at any distance; even young birds are largely silent, bar some quiet chittering noises.

Platalea leucorodia

UK STATUS Scarce migrant and increasingly resident, now breeding.

HABITAT Open, shallow coastal pools where it can swish through the water with its bill for fish and crustaceans. Nests in dense waterside trees.

Bittern

Far more likely to be heard than seen, the male's amazing foghorn boom is often the only sign of its presence in its reedbed home, and is one of the deepest and most unusual calls of any bird.

STAR SOUND: BOOMING Distinctive and easy to learn, males make very low 'booms' from about March to July, at any time of day or night but especially at dusk. It is the first sound on Track 8. It is a call that has given rise to the country names of 'Bog-thumper' and 'Bull-of-the-Bog'. He gives about four to six booms in series, each a second or so apart, followed by several minutes of silence before he calls again. The sound is like blowing air over a giant empty milk bottle. If close enough, you can hear him gulp in air several times before booming, inflating his oesophagus into a 'sounding bag'. The booms carry a mile or more in good conditions. Each male has a unique boom, some more a 'harumph' than a 'boom'.

FLIGHT CALL Birds give an occasional loud *KOW*, especially on take-off and in flight, and most often heard in autumn.

Botaurus stellaris

UK STATUS Resident, with now more than 150 booming males, up from just 11 in 1997, primarily in Somerset and East Anglia, with some in north-east England and a few other sites. Around 600 individuals winter, including some from continental Europe.

HABITAT Breeds in large, freshwater reedbeds, with shallow pools rich in fish. Can visit quite small patches of reeds in winter, often at traditional sites.

WHERE TO HEAR Ham Wall and surrounding reedbeds (Somerset); Minsmere, North Warren, Lakenheath and Walberswick (Suffolk); Cley and Titchwell (Norfolk).

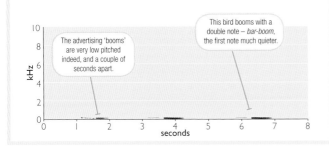

The advertising 'booms' are very low pitched indeed, and a couple of seconds apart.

This bird booms with a double note – *bar-boom*, the first note much quieter.

Little Bittern

This diminutive heron is starting to breed on the Somerset Levels. Knowing its call can be an important means of locating one, even if you never see it.

ADVERTISING CALL (MALE) At dawn and dusk gives a guttural *grrō*, repeated every 2 seconds.

FLIGHT CALL A sharp *kwak*, rather frog-like.

Ixobrychus minutus

UK STATUS Very rare summer visitor and breeder.

HABITAT Extensive reedbeds.

Grey Heron

This metre-tall, stately grey-and-white fisherman poses few identification challenges by sight, the adult especially with its long yellow dagger bill and white head with bold black 'bandana', while in lumbering flight it hunches its neck back onto its shoulders. It is not very vocal, except around the nest, and a visit to a colony is quite an experience, but its calls are still worth knowing to alert you to birds flying overhead.

FRANK! CALL The main call, often given in flight, is a very loud, harsh, raspy *FRANK!* It is often inflected upwards, unlike the straight-pitched irritation call of the Little Egret. It is frequently heard at dusk and dawn as they head to and from their roosts, and it has real resonance, especially when heard over the water.

HERONRY CALLS An active heronry can be a lively place, the noise waxing and waning as adults arrive and leave. Early in the season, males attract females with loud, deep, bellowing or trumpeting *RRŌ!* calls full of resonance, and also a higher *RAK!*, plus all manner of grunts and variants on the *FRANK!* call, sounding like a cage of wild animals! They sometimes also clap their bills, making 'chop' noises. By mid-spring, the throaty begging calls of the young become prominent, *ak-ak-ak-ak-ak....*

COMPARE WITH Little Egret; Cormorant.

Ardea cinerea

UK STATUS Widespread resident with about 13,000 pairs, mainly in lowlands and including some urban areas. Populations are hit hard by freezing weather.

HABITAT Lakes, marshes, rivers and estuaries. Breeds colonially in heronries, usually in tall waterside trees but sometimes in woods quite far from water.

WHERE TO HEAR Accessible heronries include Regent's Park and Battersea Park (London) and the heronry hide at Swell Wood (Somerset). Also try Middleton Lakes (Staffordshire), Attenborough Nature Reserve (Nottinghamshire), Ellesmere (Shropshire) and Washington (Tyneside).

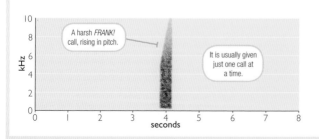

A harsh *FRANK!* call, rising in pitch.

It is usually given just one call at a time.

Little Egret

This recent arrival, a half-sized, all-white heron with black legs and yellow feet, is now a familiar sight on estuaries and wetlands despite having only bred for the first time here in 1996. However, it is at their breeding colonies that they turn really noisy, making a full range of raucous barks, growls and a wonderful *wobblywob* call.

STAR SOUND: *WOBBLYWOB* CALL Very vocal and loud at the nest and when settling into a roost, the male's most prominent advertising call stands out in the din of a colony, sounding like an exaggerated, deep, loud gargle: *WOBBLYWOB*.

OTHER CALLS AT NEST Pairs make a wide range of barks and growls, including *chak-chak-chak…* and *szchnap! szchnap!*

IRRITATION CALL When disturbed into flight or when pushed from a prime fishing spot by a rival, birds make a harsh, very bad-tempered, straight-pitched *arrrrgh*!

COMPARE WITH Cormorant; Grey Heron.

Egretta garzetta

UK STATUS Resident and increasing throughout lowland England and Wales from Lancashire southwards, with more than 1,000 breeding pairs.

HABITAT Fishes in the shallows of estuaries, saltmarshes, rock pools and some gravel pits and reservoirs. Tends to breed alongside Grey Herons in woods.

WHERE TO HEAR Swell Wood heronry hide (Somerset); Heronry Viewpoint at Northward Hill (Kent); Burton Mere Wetlands (Cheshire).

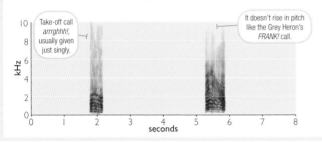

Take-off call *arrrghhh!*, usually given just singly.

It doesn't rise in pitch like the Grey Heron's *FRANK!* call.

Gannet

This impressive seabird, white with golden head feathers and black wing tips, is only really vocal at its few immense colonies on islands and cliff faces.

NEST-SITE CALL As pairs reunite or return to their young or squabble with neighbours, they give a repeated *brrrō-brrrō-brrrō…*, deep and guttural, about three a second, with a strong rolling of the 'r's. The calls rise and fall in intensity depending on the level of excitement or irritation. A colony creates a considerable chorus. It is similar to the Cormorant in pitch but is a more consistent sound, without the variety of wobbling ripples. It can sometimes give similar calls in the excitement of a mass fishing flock but is otherwise largely silent.

COMPARE WITH Cormorant; Shag.

Morus bassanus

UK STATUS 220,000 breeding pairs in just 22 gannetries, most in Scotland. In winter, they wander at sea.

HABITAT Exclusively marine, often in inshore waters, and nesting on rocky islands and cliffs.

Great White Egret

This large, all-white, yellow-billed heron has spread dramatically across western Europe and is just becoming firmly established here in the UK.

FLIGHT CALLS A rippling, deep, wooden rattle, *brrrrrr*, often quite drawn out, but sometimes faster and shorter. It can also give goose-like trumpet calls.

Ardea alba

UK STATUS Rare visitor, first bred 2012.

HABITAT Breeds in reedbeds; feeds in shallow pools.

Shag

Like a small, wholly marine Cormorant with a tufty crest, Shags tend to nest in scattered colonies, so it hasn't developed the range of calls of its more sociable cousin. You need to be up close to hear their infrequent and quiet calls among the din of other seabirds.

NEST-SITE CALLS Even though the males' calls are quite loud, they are mainly made at the nest and are rather sparing, a limited repertoire of simple, hollow-sounding, deep guttural creaks and bass grunts, usually in ponderous series, interspersed with short, throaty clicks. Females tend to just click in reply.

Phalacrocorax aristotelis

UK STATUS About 27,000 breeding pairs, most around north and west coasts. Populations disperse locally in winter.

HABITAT Exclusively coastal, breeding on cliffs and islands and feeding at sea.

Cormorant

Stood on a buoy or seastack, wings outstretched, this large, black water bird looks heraldic, prehistoric almost, but to hear it requires a visit to its nesting colonies ('rookeries').

NEST-SITE CALLS Has a wide repertoire of deep gurgling, barking and guttural chunterings, sounds that clearly come from a big, hollow, echo-filled throat. Some calls are a short ripple, there is often a nasal quality and many are a run of decelerating guffaws. The calls often include an element of vibrato, unlike the straight barks typical of the Grey Heron but not as wildly gargling as the *wobblywob* of the Little Egret.

Phalacrocorax carbo

UK STATUS About 9,000 pairs breed, with a winter population of more than 40,000 individuals.

HABITAT Predominantly coastal but also reservoirs, gravel pits and larger rivers. It nests on cliffs and on lake islands in trees or occasionally on the ground.

WHERE TO HEAR Breeding colonies include: Dungeness (Kent); Rye Harbour (East Sussex); Walthamstow Wetlands (London); Abberton Reservoir (Essex); and Paxton Pits (Cambridgeshire).

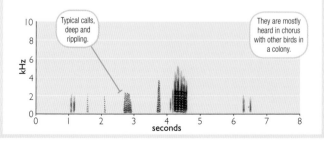

Typical calls, deep and rippling.

They are mostly heard in chorus with other birds in a colony.

Osprey

This majestic white-and-black bird of prey, well known for its fish-catching prowess, is a conservation success story following its return in 1954, having almost become extinct in the UK in the early twentieth century. It has an unexpected wader-like whistle for such a large bird.

DISPLAY CALL Early in the season, males climb high in the sky and then display, doing a series of vertical plummets and pulling out each time in a tight curve. During the performance, they call *weep, weep, weep*, not unlike the Oystercatcher, and usually strongly inflected upwards.

ALARM CALL Both sexes give a series of high blasts of a shrill whistle, *pew! pew! pew!...*, either slow at about one a second, or fast at up to five a second. Often given from the nest or nearby, it is usually quite pure, but can sometimes be slightly hoarser.

COMPARE WITH Oystercatcher.

Pandion haliaetus

UK STATUS Summer visitor, with more than 200 pairs mainly in Scotland, but small numbers now in Poole Harbour, northern England, north Wales and Rutland Water. Also a passage migrant to reservoirs and south coast estuaries.

HABITAT Needs rich fishing grounds, be that lakes, rivers, estuaries, sea lochs or fish farms! Each pair builds a giant stick nest nearby in a prominent tree.

WHERE TO HEAR Cors Dyfi (Powys); Rutland Water (Rutland); Bassenthwaite Lake (Cumbria); Loch of the Lowes (Perth and Kinross); Loch Garten (Highland).

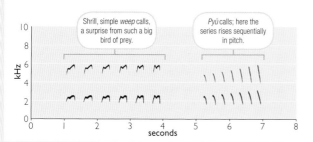

Shrill, simple *weep* calls, a surprise from such a big bird of prey.

Pyū calls; here the series rises sequentially in pitch.

Honey-buzzard

This is an elusive and rare summer visitor from Africa, not arriving until May and most obvious when flying high above its territory, soaring in tight circles or in big-dipper display flight. Despite being only distantly related, it looks very similar to the Buzzard, best told in flight by its smaller yet protruding head and subtle differences in wing and tail. This is a vocally reserved bird, not calling in display, and only really vocal if disturbed at the nest when it has chicks.

CALL The main call is similar to the simple call of the Red Kite, a high, whistled, rise-and-fall *pʸū* or longer *pee-yoo*, often breaking in yodel both on the way up and down, *p'ee-y'oo*. It doesn't have the yo-yo ending of the long Red Kite call, and is given sparingly.

COMPARE WITH Red Kite.

Pernis apivorus

UK STATUS Summer visitor, arriving in May, with fewer than 50 pairs spread very thinly mainly in southern England but also East Anglia, Wales, Yorkshire and Scotland.

HABITAT Extensive woodlands such as the New Forest, where it can hunt out wasp and wild bee nests.

Golden Eagle

Surprisingly for such a large and magnificent bird, this is perhaps one of Britain's quietest, almost never heard even by those intensely studying them. Instead of using sound to communicate, they fly high over their large territories, using their excellent eyesight to see any rival or find a mate. Even at the nest or in encounters with other eagles, they are largely silent, so finding and identifying them is by eye, where the long wings and tail and powerful flight are often the best clues when they are distant.

CALLS Rare calls include short, rather high and shrill yapping notes, $pyop$ or $kyek$, sharply dropping in pitch, including from youngsters begging for food from their parents at the nest.

Aquila chrysaetos

UK STATUS Around 440 pairs breed in wild and remote parts of Scotland, with the English and Northern Irish population now extinct and a reintroduction underway in Ireland.

HABITAT Generally found in the wildest mountains, hunting out over open habitats.

Sparrowhawk

DDT pesticide poisoning in the 1950s and 1960s devastated the Sparrowhawk population, which took several decades to recover. It is the bird of prey most likely to be seen in gardens. Adult males are bluish-grey above with rusty cheeks, the females and youngsters browner. They hunt small birds in low-level aerial ambush, and also fly higher as they scout for prey or display silently in slow, soaring circles. However, it is rarely heard except close to the nest where it has two main call types.

CHICKERING CALL A fast *chik-chik-chik-chik-chik-chik...*, about seven notes a second, given early in the breeding season and mainly from a hidden perch.

WAILING CALL Like a high-pitched version of the Buzzard's 'mew', it is also shorter and rather like a squeaky toy, with a short-rise-then-long-fall arc, $w^{e}ee$. At times, it can be upwards inflected. Begging chicks make a similar noise in late summer, sometimes incessantly, but often breaking into a stammer, $w^{e'}e'e'e'$.

COMPARE WITH Goshawk; Buzzard; Lesser Spotted Woodpecker.

Accipiter nisus

UK STATUS Widespread resident with about 35,000 pairs, commonest in the lowlands, and typically sedentary but some come from northern Europe for the winter.

HABITAT Maintains a large hunting territory, any area with woods and copses where there are plentiful small birds to catch, including urban areas.

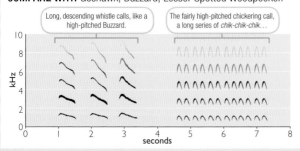

Long, descending whistle calls, like a high-pitched Buzzard.

The fairly high-pitched chickering call, a long series of *chik-chik-chik...*

kHz

seconds

Goshawk

This powerful hawk is incredibly elusive, most often seen on fine, warm spring mornings when they display above their territories. Females are Buzzard-sized, but males are little bigger than a female Sparrowhawk. They are not especially vocal birds.

CALLS Their repertoire is very similar to that of the Sparrowhawk but deeper and meatier, as befits a much bulkier bird. It is mainly heard early morning at the start of the breeding season, females in particular sitting in the trees and 'chickering' loudly, *chik-ik-ik-ik-ik-ik-ik...*, about five to six notes a second.

COMPARE WITH Sparrowhawk; Buzzard.

Accipiter gentilis

UK STATUS Resident and widespread, with about 600 pairs.

HABITAT Extensive conifer and mixed forests, hunting within the trees and over open ground nearby.

WHERE TO HEAR New Forest (Hampshire); New Fancy Viewpoint (Forest of Dean, Gloucestershire); Wykeham Forest Raptor Viewpoint (North Yorkshire).

Marsh Harrier

From just one pair in 1971, this species is fortunately now a familiar sight over many wetland nature reserves. The most distinctive call is from 'sky dancing' males.

DISPLAY CALL (MALE) In early spring (January–April), males fly high over the nest site and then plunge-dive, doing extraordinary acrobatics while making a short, squeaky *we^e^o* call, often calling once or twice followed by a pause.

WICKERING CALL In alarm, both sexes give a range of fast *chikikikikik...* calls, up to 10 or so notes a second.

Circus aeruginosus

UK STATUS More than 400 'pairs' (many males support more than one female), mainly in eastern England from Kent to Yorkshire, some resident, some migrating south for the winter.

HABITAT Breeds in large reedbeds but also in arable crops. It also hunts over coastal marshes, looking for small birds and mammals.

WHERE TO HEAR Great sites include: Lakenheath and Minsmere (Suffolk); Cley and Titchwell Marsh (Norfolk); and Blacktoft Sands (East Yorkshire).

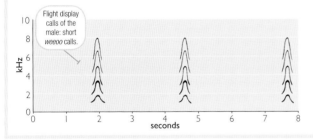

Flight display calls of the male: short *weeoo* calls.

Montagu's Harrier

Only around 10 pairs of this elegant summer visitor breed in the UK. Their appearance is reminiscent of a slim Hen Harrier.

CALLS A very similar and just as limited repertoire as the Hen Harrier, but slightly higher pitched; males sometimes give a *chek chek chek* call, and both sexes having 'wickering' calls if alarmed.

Circus pygargus

UK STATUS Very rare summer visitor.

HABITAT Cereal fields on downland or coastal plains.

Hen Harrier

On a few windswept heather moors, the ghostly grey males make a dramatic sight in spring when sky dancing, but neither theirs nor the females' calls are often heard.

DISPLAY CALL (MALE) Occasionally males give a straight, chattering *chuk-uk-uk-uk*, typically quite short, sometimes given on ascent in the sky dance; otherwise largely silent.

ALARM CALL If really agitated, it can give a fast 'wickering': *wik-ik-ik-ik-ik…*, typically from the female at the nest, about 10 notes a second.

COMPARE WITH Marsh Harrier; Montagu's Harrier.

Circus cyaneus

UK STATUS Rare resident, still persecuted (illegally) in areas; in 2016, there were 460 breeding pairs in Scotland, 46 in Northern Ireland, 35 in Wales, 30 on the Isle of Man, but just a handful in England.

HABITAT Breeds on upland heather moors, moving in winter onto coastal marshes, fens, downland and lowland heathland.

Red Kite

Intense persecution in the nineteenth century almost drove this graceful raptor to extinction. It hung on in mid-Wales until a reintroduction scheme began in 1989 in several parts of the country. The population has since soared and continues to climb, allowing many people to now enjoy it as it drifts languidly overhead, its forked tail twisting constantly in flight like a rudder, and also to hear its human-like whistle calls.

WHISTLE CALL The far-carrying calls are most likely to be heard around the nest site and at feeding stations, often given by a perched bird in spring and used by a pair to keep in contact. It is a simple, high, rather long whistle, $p^{eeooooo}$, almost like a human whistle in feel rather than the Buzzard's cat 'mew'. The call whips up in pitch at the start, sometimes with a slight yodel-break, and then drifts down. It will often add extra, faster notes at the end, yo-yoing up and down in a fun way, such as $p^{eeooooooo}$ $w^{ee}oo$-$w^{ee}oo$-$w^{ee}oo$.

COMPARE WITH Common Buzzard; Honey-buzzard.

Milvus milvus

UK STATUS Largely resident, although many first-year birds wander widely in spring. Around 2,500 pairs now breed, across most of Wales and in broad clusters around the nine release sites in England, Scotland and Northern Ireland.

HABITAT Hunts by leisurely circling over open farmland, heaths, parkland and roadsides. Nests in woodland and copses, and roosts communally in winter.

WHERE TO HEAR The Chilterns (especially in Bucks); Cwm Clydach (Swansea); Gwenffrwd-Dinas (Carmarthenshire); Gigrin Farm (Powys); Ynys-hir (Ceredigion); Top Lodge (Northamptonshire); Harewood House (West Yorkshire); Galloway Red Kite Trail (Dumfries and Galloway).

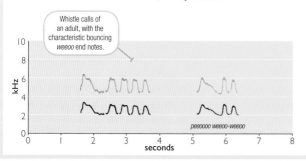

Whistle calls of an adult, with the characteristic bouncing *weeoo* end notes.

peeoooo weeoo-weeoo

White-tailed Eagle

Famously described as a 'flying barn door', this giant bird is one of our most impressive, appearing to move in slow motion yet manoeuvrable enough to power-grab large fish from the water in flight. Most birds of prey aren't especially vocal, yet this species bucks the trend with pairs often duetting loudly.

CALL Most vocal at and near the nest site in the breeding season, especially at dawn, adults make a loud yapping noise, like a small lapdog, *AK-AK-AK-AK-AK-AK...*, about five a second, and for 10 seconds or more. The sound reverberates around the wild scenery.

DUET A pair will often 'yap' in duet, their calls becoming synchronous, the female's call a full octave lower than the male's. This duet can be given side by side at the nest, or with the two birds perched a considerable distance apart. At its most impressive, the birds will fling their heads back while calling, just as a wolf howls to the sky.

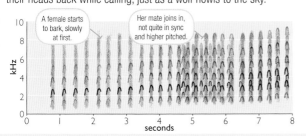

A female starts to bark, slowly at first.

Her mate joins in, not quite in sync and higher pitched.

Haliaeetus albicilla

UK STATUS Persecuted to extinction in the UK by 1918; a Scottish reintroduction programme has been underway since 1975 and the number of breeding pairs now exceeds 100, mainly along the western Scottish seaboard but with a reintroduction programme underway in north-east Scotland.

HABITAT Mainly wild, undisturbed coastal areas, fishing in the sea close to shore. Usually nests in large trees, maintaining stable territories over many years.

WHERE TO HEAR Mull (Argyll and Bute), Skye and Wester Ross (both Highland) remain the core of their range. Special guided trips allow you access to hides overlooking nest sites.

Buzzard

Track 65

After a history of persecution and pesticide poisoning, this is now the nation's most numerous raptor, seen almost everywhere on roadside posts or soaring the thermals, with a mewing call that is one of the evocative sounds of the countryside. See page 100 for full description.

'MEW' CALL The long *myewwwwww* is more a catcall than a human whistle. It is almost always drawn out, lasting a second or so, with a long, sliding fall in pitch, sometimes wavering a little at the end. It is given mainly when high-soaring, but can be given from trees. Juveniles often give repeated, higher-pitched wails, flatter in pitch for the main core of the call before falling at the end.

ALARM CALL In aggression or when nervous, the call becomes faster, more intense and shrill, and can waver towards the end, but without the yo-yoing of the Red Kite call.

COMPARE WITH Jay; Red Kite; Honey-buzzard; Sparrowhawk; Goshawk.

Buteo buteo

UK STATUS Sedentary resident across much of the UK, with around 70,000 pairs.

HABITAT Mixed farmland with woods and copses, upland or lowland, but rare in urban areas. Although a skilled hunter of Rabbits, it will also happily sit eating worms on a ploughed field.

Water Rail

Rare glimpses, often along a reedbed margin at dusk, are of what looks like a thin Moorhen with a long, downcurved red bill, but this is a bird much more often heard, for this is the 'piglet of the reedbed' with wonderfully squealy calls.

STAR SOUND: SHARMING CALL This call, with its own special name, is especially used when two rivals meet. It sounds like a fractious piglet. Usually several long notes are given in series, each starting with a deep, low snort, breaking in pitch to a full, squealy scream: *schnWEEEER*. The calls increase in intensity to a peak of indignation, and then calm and slow to a deep, grumpy grumble. Bass notes rumble underneath like a hungry stomach.

ADVERTISING CALL (MALE) In spring, the male gives a simple, sharp *plik!* call, like a sculptor tapping slowly with a chisel on a rock. It is repeated, often incessantly for a minute or more, at a rate of about one to two a second, like the *spink!* of the Coot but deeper in pitch.

ADVERTISING CALL (FEMALE) To attract a male, she calls *pick pick plu'eerrrrr*, with a vibrato ending.

COMPARE WITH Moorhen; Coot.

Rallus aquaticus

UK STATUS Sedentary resident, scattered widely across lowland Britain, with between 1,500 and 6,000 pairs – exact numbers are hard to judge. More arrive from the continent for the winter.

HABITAT Largely found in freshwater reedbeds and marshes with standing water, especially with a diversity of cover, but in winter also turns up in smaller wetlands, even wet ditches.

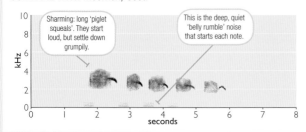

Sharming: long 'piglet squeals'. They start loud, but settle down grumpily.

This is the deep, quiet 'belly rumble' noise that starts each note.

Corncrake

The unmistakable double-rasp advertising call is all you normally experience of this secretive and rare land-rail. Once abundant in hay meadows, your best bet to hear it is now north-west Scotland.

STAR SOUND: ADVERTISING CALL (MALE) The scientific name is all you need to remember, for this 'driest' of bird sounds is just a harsh rasp, uttered twice, followed by a pause: *crex crex* (pause)…, repeated about once a second for long periods, the sound carrying far. It is given from May to July, especially at dawn and dusk, and at times through the night. Males call from dense ground cover but might be glimpsed, head raised, beak wide open.

Crex crex

UK STATUS Summer visitor with around 1,300 pairs, mainly in the Outer Hebrides, Orkney and Ireland.

HABITAT Lush damp areas on the machair (coastal hay meadows).

Spotted Crake

Like a rather small, brown Moorhen densely flecked with white, this is a supremely elusive bird that can creep unseen in the shortest of waterlogged vegetation.

ADVERTISING CALL In the breeding season, in a very few of our wildest marshes, males attract females with a very consistent call like a sharp whiplash, *hwitt!*, carrying a kilometre or more. It is mainly uttered from dusk to dawn, often in long bursts at a steady rate of about one a second, sometimes with a female quietly duetting in instant reply.

Porzana porzana

UK STATUS Fewer than 30 pairs breed in the UK and a few are seen elsewhere on spring and autumn migration.

HABITAT Tussocky wet grassland and sedge beds, visiting some reedbeds on migration.

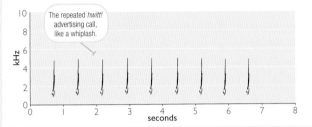

The repeated *hwitt!* advertising call, like a whiplash.

Moorhen

High-stepping along the water's edge or sculling jerkily across the water, this is a familiar water bird, but in places it can be a skulker so its calls can give away its presence.

BROOK CALL This territorial call is an explosive *BRROOK!*, rich in 'r's, given without warning and only sporadically, usually from within cover.

KI-KECK CALL This advertising call is quite abrupt and loud, a quick *ki-kek!* or *ki-tik*, the first note higher than the second.

ALARM CALL Often delivered slowly in an uneven series, each *kuk* note is rather short and sharp like the Coot's eponymous call.

OTHER CALLS It makes a wide range of other calls, some rather squealy, some a gentle clucking.

COMPARE WITH Coot ('coot' call); Water Rail (advertising call).

Gallinula chloropus

UK STATUS Resident and sedentary, common in the lowlands but increasingly scarce at higher altitudes. It has a population of more than 270,000 pairs.

HABITAT The 'moor' in its name refers to lowland fens rather than upland heather moors, and it can be found on large ponds, canals, lakes and slower rivers, wherever there is abundant waterside vegetation.

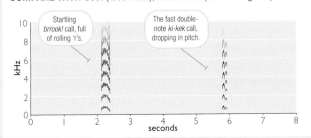

Startling *brrook!* call, full of rolling 'r's.

The fast double-note *ki-kek* call, dropping in pitch.

Coot

Sometimes mistaken for a Moorhen, the Coot can be distinguished by its white frontal shield, ivory bill and habit of diving in deep water.

'COOT' CALL The origin of the bird's name, this call has many subtle variations, but all are short notes, some trumpeting, some more raspy. It can sound like 'coot', *kyūt!*, *kroot!*, or a less intense *pook*. It is often repeated, about one a second, sometimes for several minutes.

SPINK! CALL A very brief, sharp *spink!* or *tink*, like a tiny hammer hitting marble, usually in slow series, and higher pitched than the Water Rail's advertising call. It is the fourth sound on Track 5.

FLIGHT CALL In flight at night, it gives slurred notes, like weak wavering toots on a toy trumpet.

OTHER SOUNDS It makes loud splashing noises when kickboxing an opponent or when running across the water.

COMPARE WITH Moorhen; Water Rail (advertising call).

Fulica atra

UK STATUS Sedentary resident, with more than 30,000 breeding pairs spread widely across mainly lowland areas. Also a winter visitor from eastern Europe and Russia, boosting the winter population to nearly 200,000 birds, when it often gathers in large flocks.

HABITAT Gravel pits, lowland lakes and reservoirs, preferring those with plenty of marginal and especially submerged vegetation, and enjoying the chance to come out onto short bankside grass to feed.

Crane

The triumphant duet and flight calls of this stately bird reverberate in spring around a few special marshlands.

STAR SOUND: DUET The calls have a jubilant quality, full of rich, brassy resonance, with a bold rolling of 'r's quickly breaking into full trumpet blast: *BRRŌHH!* Usually given in duet, either the male or female initiates and their mate instantly replies; the male's call is higher, giving the effect of *BRRŌH BRROW* or vice versa. It is the second sound on Track 7.

FLIGHT CALL A combination of the *BRRŌH* calls with other varied trumpets and goose-like calls, including ones with a rich ripple. Youngsters in autumn have a high-pitched sibilant *seep*.

COMPARE WITH Whooper Swan; Bewick's Swan.

Grus grus

UK STATUS Sedentary resident, until recently mainly in the Norfolk Broads, now expanding into the Fens and Yorkshire; a reintroduction scheme began in Somerset in 2010. The total breeding population was about 20 pairs in 2017 and is increasing slowly.

HABITAT Cranes need large home ranges with extensive reedbeds for nesting, arable fields and grassy marshland for feeding, and freedom from disturbance.

WHERE TO HEAR Somerset Levels; Slimbridge (Gloucestershire); Lakenheath (Suffolk); Hickling Broad and Horsey Gap area (Norfolk).

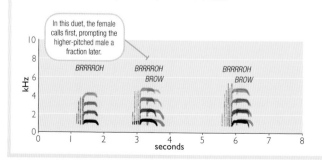

In this duet, the female calls first, prompting the higher-pitched male a fraction later.

BRRRROH | BRRRROH BROW | BRRRROH BROW

Stone-curlew

Largely nocturnal, this curious bird seeks out the driest places. At night, the powerful, wailing calls are eerily magnificent.

STAR SOUND: ADVERTISING CALL Vocal and loud, at nightfall a pair bursts into a loud, excited ku-$rrrr$-loo or plu-rrr-wee. The middle syllable is often a buzzy trill. The calls are repeated, building in volume and intensity, then calming down. This is given sporadically through the night.

OTHER CALLS The wide repertoire includes a rising wu-ee or wur-lee or Oystercatcher-like shrill wu-eep. It can start with a buzzy trill, $wrrr$-wip, or with rapid notes, $p'p'p$-eee.

COMPARE WITH Curlew; Oystercatcher.

Burhinus oedicnemus

UK STATUS Summer visitor from southern Europe and north-west Africa, arriving as early as March and some lingering until November, with around 400 pairs in East Anglia and on the downs and plains of central southern England.

HABITAT The driest, sandiest acid grasslands and flint-strewn fields, with short vegetation.

WHERE TO HEAR Minsmere (Suffolk); Weeting Heath (Norfolk).

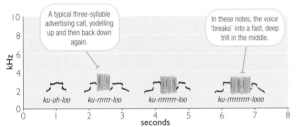

Oystercatcher

This smart, pied wader with a long red bill is a very loud and excitable presence, both on the coast and, in summer, in the uplands.

FLIGHT CALL Loud, ringing, bold and confident, it is a straight-pitched *WEEP!* or shorter *WIP!*, or with a short introductory note, *w'WEEP!*

STAR SOUND: PIPING DISPLAY Given by a single bird or pair in response to another flying over, or by rival pairs clashing where their territories meet, they charge about, shoulder to shoulder, beaks down, making a right fuss. The call usually starts with well-separated *wip* calls but accelerates, builds in volume and then breaks into bubbling trills.

COMPARE WITH Avocet; Stone-curlew.

Haematopus ostralegus

UK STATUS Common breeding visitor with a population of about 110,000 pairs, most abundant from the Pennines northwards. In winter, they are joined by others from Iceland, Norway and the Netherlands, with a total population of about 340,000 individuals.

HABITAT Breeds on moorland edge, upland arable fields, river valleys and shingle beaches, with small numbers nesting inland on gravel pits. In winter, it moves to both muddy and rocky coasts.

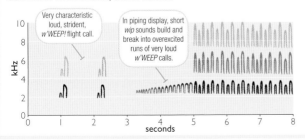

Avocet

The Dutch call this elegant wading bird *Kluut*, all you need to remember its calls, which dominate many a coastal wetland in spring.

KLŪT CALL Very vocal at breeding sites, it uses one basic call, *klūt*, but at various levels of intensity. At times it sounds more like *kleep* or *kūp*, and could be confused with an Oystercatcher. It is piping and insistent, rising sharply in pitch to a shrill end. It can be given singly but is at its most intense when flying up to bravely fend off aerial predators, when it is repeated 2–3 times a second and sounds rather knickers in a twist!

COMPARE WITH Oystercatcher.

Recurvirostra avosetta

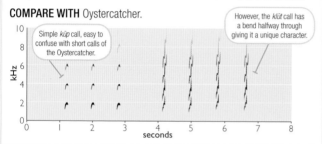

Simple *kūp* call, easy to confuse with short calls of the Oystercatcher.

However, the *klūt* call has a bend halfway through giving it a unique character.

UK STATUS Around 1,700 pairs breed in colonies at more than 100 sites. Some move south and west to winter in Europe and Africa, but about 7,500 birds remain in the UK.

HABITAT Breeds on bare islands on shallow pools, often brackish or saline. In winter, moves mainly onto sheltered, muddy estuaries.

Lapwing

The country name of 'Peewit' evokes the gorgeous calls of this plover, with the full flight display call a wonderful elaboration on this theme.

STAR SOUND: DISPLAY CALL AND WING NOISE As the male swoops in display flight, he calls: *peeeee idl-wit i-wit i-wit pee-er-wit*. The *pee* noises have a 'comb and paper' buzz to them, while the *i-wits* whip wildly in pitch. The wing tips create a pulsating *whup whup whup*. It is the fourth sound on Track 9.

OTHER CALLS Contact calls include a rising *pee-eee*, and a falling *pee-erwit* and *pee-wup*; all have an aching, desperate feel. When their chicks are threatened, the call from agitated adults gets even more panicky and intense.

COMPARE WITH Black-tailed Godwit.

Vanellus vanellus

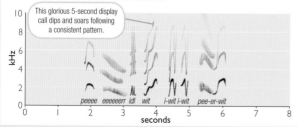

This glorious 5-second display call dips and soars following a consistent pattern.

peeee eeeeeerr idl   wit   i-wit i-wit   pee-er-wit

UK STATUS Widespread but declining breeding bird, with around 130,000 pairs, scarce in south-west England, Wales and Northern Ireland. The winter population is around 650,000 birds including some continental birds.

HABITAT Breeds in a range of open habitats, including wetlands, farmland and moorland. In winter, flocks gather on wet grassland and estuaries.

Grey Plover

Calls of this silvery-grey wintering wader with black armpits are easy to overlook on its estuary haunts, but they are subtly distinctive and very useful once learnt.

FLIGHT CALL A simple, slurred *plee-oo-wee* that drops in pitch and then bends back again, as if the sound warps.

COMPARE WITH Golden Plover.

Pluvialis squatarola

UK STATUS Winter visitor and passage migrant, with around 43,000 individuals. A few non-breeding young birds stay here during the summer.

HABITAT Muddy estuaries.

Golden Plover

The far-carrying, plaintive display calls of this golden-spangled, black-bellied wader are a characteristic sound of upland moors in spring.

DISPLAY CALL At dawn and dusk in spring, males circle high above their territories and give a drawn-out *per PEEEE-yō* in slow series. Each call lasts almost a second followed by a gap of about the same length. Every now and then it switches to a faster, neatly rhythmic *p'prrr-ye p'prrr-ye*, especially when coming down to land.

FLIGHT CALL A rather vague *pew* or *pū-ee*, almost straight in pitch but with slight warping of the sound, up or down.

ALARM CALL Vigilant birds will signal danger with repeated *too* calls.

COMPARE WITH Grey Plover (call).

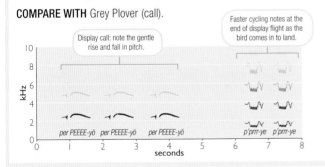

Display call: note the gentle rise and fall in pitch.

Faster cycling notes at the end of display flight as the bird comes in to land.

per PEEEE-yō   per PEEEE-yō   per PEEEE-yō   p'prrr-ye   p'prrr-ye

kHz / seconds

Pluvialis apricaria

UK STATUS Summer visitor to upland and northern breeding grounds, with around 50,000 pairs. In winter they are joined by birds from northern Europe, to give a total population of about 340,000.

HABITAT Breeds on heather moorland. Winters on grazing marshes, estuaries and tilled fields in large flocks, often with Lapwings.

Dotterel

This rare wader of the wildest Highland mountains has rather soft calls.

DISPLAY CALL The female in display flight gives a simple series of *pwip* calls, up to four a second. It is also given singly as a contact call.

TAKE-OFF CALL A deeper, sweetly buzzing *pyurrrr*, downwards inflected.

COMPARE WITH Golden Plover (call).

Charadrius morinellus

UK STATUS Rare summer visitor to northern Scotland, with around 400 breeding males.

HABITAT A specialist of tundra-like high mountain plateaus.

Ringed Plover

Moving in a 'pause-and-dart' way on drier mud, the two 'ringed plovers' are very similar in appearance. The differences in their flight calls and display calls can greatly aid identification. The orange legs and white wing bar separate this more coastal, year-round species from the Little Ringed Plover, but such differences are difficult to make out at any distance.

DISPLAY CALL In spring, it gives two forms of display: in the threat display, rival males scuttle towards each other, hunched up, tail fanned and lowered; in flight display, males zigzag low over their territories in so-called 'butterfly' flight. In both, they make constant, fast, cyclical calls in long series, $t^{l}ew$-a 't^{l}ew-a' $t^{l}ew$-a..., or $t^{l}lew$ $t'lew$ $t'lew$..., in which the *lew* is often a lower, strangely squelchy note.

FLIGHT CALL Given at any time of year, it is a disyllabic $p\bar{u}$-ip, with a slight rise in pitch. Flocks can also give softer, short *wip* notes.

Charadrius hiaticula

UK STATUS About 5,400 pairs breed, with the largest populations on the Scottish islands. About 36,000 birds winter here, and migrants from the Arctic pass through in spring and autumn.

HABITAT Breeds on shingle and sandy beaches and increasingly on gravel pits inland.

Little Ringed Plover

This delightful little summer visitor with its yellow eye-ring and dark legs is often known as the 'LRP', and bred here for the first time in 1938.

FLIGHT CALL Typically, the call is a slightly but perceptibly downslurred $p\bar{u}$-yip, although in alarm the pitch is straighter.

DISPLAY CALL The manner of the flight display is similar to the Ringed Plover, as it flickers low and fast over its territory, but the call is usually a two-syllable couplet in repeated series, two per second, a buzzy *dree-zup dree-zup dree-zup....*

THREAT CALL Directed by the male at any rival, it is a buzzing, fast-shuttling series of repeated notes, about six to eight per second and lasting for several seconds, such as *zri-zri-zri-zri-zri-zri-....* It sometimes starts slowly and accelerates before calming down again.

Charadrius dubius

UK STATUS Summer visitor from Africa with maybe 1,200 pairs in mainly lowland England, with some in Wales and eastern Scotland.

HABITAT Mainly freshwater sites, quickly colonising freshly dug gravel pits; also seen on shallow rivers with shingle bars and islands.

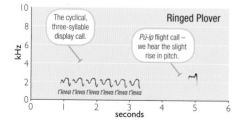

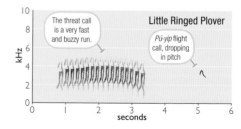

Curlew

To hear the bubbling display calls of this, our largest wader, over a moor or marsh is a real thrill. Sadly, populations are falling fast.

STAR SOUND: DISPLAY CALL This far-carrying call starts slowly with simple, long *koooor* notes that get quicker, louder and rise gently in pitch, breaking into long *kooorlee* notes with full vibrato. It is given from the ground but especially in display flight.

CALL The basic call is *koor-lee*. It is the fifth sound on Track 8. Also gives a much shorter flight call, *kew*, often in doubles or triples.

ALARM CALL A stammering, vibrato *cha-cha-cha-cham*.

COMPARE WITH Whimbrel; Stone-curlew.

Numenius arquata

UK STATUS About 68,000 breeding pairs, mainly from the Pennines northwards, most moving south-west for the winter and joined by birds from northern Europe, giving a winter population of about 150,000.

HABITAT Predominantly a breeding bird of moorland edge, but also found on lowland wet grassland and a few on wet lowland heath.

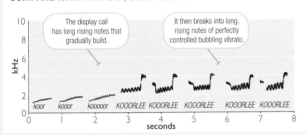

The display call has long rising notes that gradually build.

It then breaks into long, rising notes of perfectly controlled bubbling vibrato.

koor   koor   kooooor   KOOORLEE KOOORLEE   KOOORLEE   KOOORLEE KOOORLEE

Whimbrel

The calls of this stripy-headed cousin of the Curlew are heard around many coasts on spring migration. However, you must head to Orkney and Shetland for the full display call, a superb, pulsating bubbling sound.

STAR SOUND: DISPLAY CALL Rising 'wind-up' notes break into a pulsating whistled trill. The key feature is the end trill – a straight, unbroken bubble that can last 8 seconds or so.

CALL In flight, especially just before and during take-off, it gives a steady series of whistles, usually about 7–12 notes but at times up to 30: *pi-pi-pi-pi-pi-pi-pi-pi*, at around 10 notes per second.

COMPARE WITH Curlew (call and display call).

Numenius phaeopus

UK STATUS Breeding visitor to Shetland and Orkney, with only around 300 pairs. Otherwise, a passage migrant in small flocks mainly along western coasts, en route to northern Europe. A handful winter on southern estuaries.

HABITAT On migration, stops off on muddy estuaries and on the short turf of coastal fields. On the breeding grounds, uses heather moorland.

WHERE TO HEAR In the breeding season, try Fetlar and Unst (Shetland).

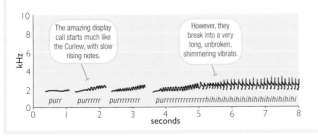

The amazing display call starts much like the Curlew, with slow rising notes.

However, they break into a very long, unbroken, shimmering vibrato.

purr   purrrrrr   purrrrrrrrr   purrrrrrrrrrrrrrrhihihihihihihihihihihihihihi

Bar-tailed Godwit

This long-billed wading bird winters on our larger estuaries, often feeding with flocks of Knot and Dunlin, with the highest numbers by far in The Wash (Norfolk/Lincolnshire). Slightly shorter legged than its Black-tailed cousin and without the bold wing bars in flight, it looks rather drab and streaky in winter. However, in early May, migrating flocks from Africa pass along the English south coast in full breeding plumage, entirely brick-red below, silver-spangled above.

CALL Its display call is reserved for its Arctic breeding grounds, so here in the UK all we get to hear are quiet contact calls, mostly a nasal double-note *che-chek*. Given that the birds are often far out on the mudflats, this can be very difficult to pick out.

Limosa lapponica

UK STATUS Winter visitor and passage migrant, with a population of just over 40,000 birds.

HABITAT Almost exclusively on coastal mudflats and estuaries.

Black-tailed Godwit

It is fairly easy to recognise that you are looking at one of the two godwit species: large, long-legged waders with long, straight(ish) bills. In flight, the strong white wing bar and black tail of the Black-tailed Godwit are obvious distinctions from its Bar-tailed cousin, but it is also more vocal (and variably so), include a *wicka wicka* call used mainly in display. If you find a noisy godwit, it is usually this one!

DISPLAY CALL Excitedly vocal in the breeding season; this variable call typically seesaws in a two-syllable ^{WICK}-a ^{WICK}-a rhythm; variants include *WIT-tew WIT-tew*....

OTHER CALLS Gives a range of single *wik* calls, double *zi-zit* notes, or a fast *wik-ik-ik-ik-ik-ik*.... Also regularly heard are more Lapwing-like notes, including a rise-and-fall *w^{ee}-erjjjjjjj*, the ending fuzzy.

COMPARE WITH Snipe; Lapwing.

Limosa limosa

UK STATUS Rare breeder, with only around 50 pairs, but a few thousand non-breeders summer here; the winter population, mainly from Iceland, is around 44,000.

HABITAT Breeds on large, wet grassland sites. In winter, flocks visit sheltered estuaries and nearby freshwater pools and meadows.

Knot

This dumpy wader gathers in large flocks out on winter mudflats, and they swirl in dramatic tightly packed formations when pushed off by the tide or spooked by a Peregrine. Seen close to, they are rather anonymous in grey winter plumage, but in late spring we see a few in their breeding finery: rufous-orange below, with silver, black and orange freckles above.

CALL They are surprisingly quiet, even en masse. The soft *wek wek* call is usually only audible at close distance.

COMPARE WITH Dunlin.

Calidris canutus

UK STATUS Passage migrant and winter visitor to our largest estuaries, especially The Wash (Norfolk/Lincolnshire), with a population of about 330,000.

HABITAT Prefers the largest estuaries and mudflats.

Turnstone

This winter-visiting beachcomber prefers rocky and shingly coasts rather than mudflats, and its 'triple ripple' contact and flight call is a useful alert to its presence. In winter plumage, it is well camouflaged among the stones, being dark brown above and on the breast, with a white belly, but in spring it moults into sublime tabby-cat black, white and orange.

CALL The cheerful call, given especially on take-off and in flight, is rather distinctive and well worth learning. It is a very fast *chi-ti-tik* or *kut-ut-uk*, on a fairly constant pitch, usually three syllables but somewhat variable, creating a pleasing, short, metallic ripple. It also gives a single-syllable *turk* note and other more conversational and quiet 'witters' and 'chitters' when feeding together.

COMPARE WITH Purple Sandpiper.

Arenaria interpres

UK STATUS Widespread around the coast in winter, although rarely in flocks of more than 50, with a winter population of just over 50,000. It leaves in spring for breeding grounds in northern Greenland and Canada; a few non-breeding birds summer in the UK.

HABITAT Rocky shores, shingle beaches, sea defences, piers and jetties.

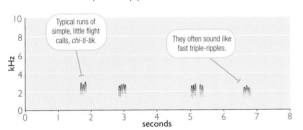

Typical runs of simple, little flight calls, *chi-ti-tik*.

They often sound like fast triple-ripples.

Ruff

Fewer than 10 females nest in the UK each year, so we have little chance of seeing the bizarre lekking dances of the larger males in their extravagant head ruffs. Instead, we see birds in dowdier non-breeding plumage, either on passage or in winter. However, in all situations they are effectively silent.

Calidris pugnax

UK STATUS Scarce passage wader, with about 800 in winter.

HABITAT Shallow wetlands.

Curlew Sandpiper

A handful pass through the UK in spring in their brick-red breeding finery. However, more are seen in autumn on their way back south to Africa, first the adults in grey winter plumage and then the buffier-coloured juveniles; the white rump can be important for identification in all plumages. It tends to be quiet when feeding, calling only on taking flight.

FLIGHT CALL A simple *chid-it* or *chir'p*, quite different from the buzzy call of the Dunlin.

Calidris ferruginea

UK STATUS Scarce passage migrant, mostly seen in autumn.

HABITAT Mainly stops off at prime freshwater lagoons along the coast with muddy margins, often associating with flocks of Dunlin.

Temminck's Stint

Every May, about a hundred of these tiny creeping waders with yellowish legs pass through the UK on their way to breeding grounds in northern Europe, rarely stopping more than a day or two.

CALL The distinctive call is a high-pitched *sirrirrit*, typically a very rapid five notes, usually given on take-off and with more than a hint of the sleigh bells of the Waxwing.

DISPLAY CALL Rarely heard in the UK, breeding males have an amazing long display call, given mostly in flight. Sometimes lasting a minute or more, it is a pulsing, grasshopper-like trill that gently wanders up and down in pitch.

COMPARE WITH Waxwing.

Calidris temminckii

UK STATUS Scarce passage migrant, mainly in spring. One or two pairs sometimes breed in remote parts of northern Scotland.

HABITAT Passage birds visit small, shallow pools, often inland, rather than typical wader haunts along the coast. Breeding birds use extensive marshlands.

Sanderling

This frosty-looking wader sprints along the tide's edge on winter beaches like a clockwork toy, its legs a blur. When flushed, it tends to fly low to quickly relocate 50m or so further on. In spring some birds that are entering breeding plumage, with black and rufous flecks showing through the white, pass through.

CALL Not very vocal; its call is a simple, rather high-pitched *plik* or *wit*, short and fairly sharp but not too loud, usually uttered when disturbed, sometimes calling several times but not in a rhythmic way. Several birds together sound a little Linnet-like.

COMPARE WITH Knot; Dunlin.

Calidris alba

UK STATUS About 17,000 birds winter here, with more passing through in spring and autumn.

HABITAT Almost wholly restricted to sandy shores.

Little Stint

This diminutive wading bird is often seen with Dunlins as they peck around hyperactively on muddy pools. The grey-plumaged adults tend to pass through first; the juveniles, which have neatly marked upperparts including white 'braces', move through rather later.

CALL If you are lucky enough to see them close to, you may sometimes hear their calls as they take flight, which are thin, high *sip*, *tip* or *chip* notes, rapid and rather feeble. Their display calls are reserved for their breeding grounds in the Arctic.

Calidris minuta

UK STATUS Passage migrant in variable numbers, with only a few seen in spring and rather more in autumn, although usually singly or in small groups. Also a regular winter visitor in very small numbers.

HABITAT Coastal lagoons with muddy fringes. Winters on southern estuaries.

Dunlin

This is the commonest and most familiar of all the small waders on winter estuaries, feeding furiously in large flocks. By spring, it moults into its smart breeding plumage, rufous above and with a big black belly patch, but you'll need to follow it to its breeding bogs to hear its pulsating display calls.

FLIGHT CALL A simple, straight-pitched *zreep* or *zrrrrt*, like a miniature referee's whistle, but you will need to be close to hear them.

STAR SOUND: DISPLAY CALLS There are two main calls, given either in a fluttering, dip-and rise display flight, or in a curious ground display in which it raises one wing vertically, like semaphore. One is a series of deep, drawn-out croaking noises, about one per second: *zwarr zwarr zwarr....* The other is a long, energy-filled trill that lasts around 4–7 seconds, like a referee's whistle as if played by a skilled musician. It starts with some long, drawn-out *zree* calls, each a little higher in pitch than the one before, building up and then cutting to a fast buzzing, pulsing note, *zizizizi…*, that gently slows, crescendos and drops in pitch through to its end.

Calidris alpina

UK STATUS Breeds mainly in Scotland, with around 10,000 pairs. Otherwise a winter visitor and passage migrant, including many birds from Iceland and some from Greenland and northern Europe, a total of around 350,000 birds.

HABITAT Breeds on high moors, especially where there are extensive bogs, and on coastal machair in the Scottish islands. In winter, mainly on muddy estuaries and coastal pools.

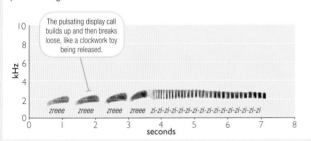

The pulsating display call builds up and then breaks loose, like a clockwork toy being released.

Purple Sandpiper

This discreet little wader, more dusky-coloured than purple and with yellowish legs and a yellow-based bill, appears to take its life in its own hands on rocky, wave-lashed shores, often returning to the same favoured locations each year. There, with luck, you may hear the calls above the sound of the sea.

CALL It is quietly conversational when feeding or in flight. It typically gives sharp little *kvit!* notes, slightly hoarse, often tumbling into pleasant little twitters and chatters, but difficult to hear above the sound of the waves.

COMPARE WITH Turnstone (call).

Calidris maritima

UK STATUS Winter visitor with about 13,000 birds, most in the northern half of Britain. Occasional pairs breed in the Scottish Highlands.

HABITAT Inhospitable rocky shores plus favoured piers and jetties.

WHERE TO HEAR Rocky coasts from Northumberland to Shetland, or try Filey Brigg and Whitby Harbour (North Yorkshire) and Hartlepool Headland (County Durham).

Woodcock

Despite being a relatively common and widespread bird, especially in winter, few people ever see a Woodcock, for it sits tight on the woodland floor by day, perfectly camouflaged as dead leaves. You may sometimes accidentally flush one in a winter woodland but it will rise into the air silently (unlike the Snipe) and soon steal away through the trees.

Instead, head out on spring evenings at dusk to extensive woods and heaths to see males flying in random criss-crossing circuits, a kind of aerial lekking system called 'roding', as they compete for the best airspace. They have a long-billed and rotund flight silhouette, with broad, arched, rather bat-like wings, but it is the strange calls that will give you a head start in finding one.

'BELCH AND SQUEAK' RODING CALLS Males are consistently and reliably vocal when roding. Every few seconds in mid-flight, one will make two or three very deep, dry, belch-like sounds, *warp! warp!*, quickly followed by a high-pitched, double-syllable *piz'p!* The noises are not loud or attention grabbing, yet they travel surprisingly far in the night air, and once learnt will alert you to a bird before it comes into view.

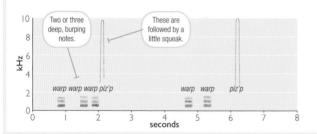

Scolopax rusticola

UK STATUS Resident with about 75,000 breeding females thinly spread across the country, scarcest in Wales and the English south-west and Midlands. It has declined significantly since the 1960s. In winter, the population may be a staggering 1.4 million birds, boosted by migrants from northern Europe.

HABITAT In the breeding season, it is found in extensive, open and often healthy woodland, including young plantations. In winter, it is widespread in woods. At all times, it needs damp ground for nighttime feeding.

Jack Snipe

Looking very much like a rather small and short-billed Snipe, but with a dark crown and 'tiger-striped' back, it sits tight in boggy grassland and saltmarsh, remaining unseen unless almost stepped on, when it will spring into the air. Occasionally, a bird may emerge into view on the edge of a marsh, where it will often indulge in its curious habit of bobbing up and down constantly, as if on springs.

CALL Unlike the Snipe, it rarely calls when flushed, but it can make a similar *skatch!* call, so be aware that a 'snipe' that calls isn't always the Snipe. Its strange 'galloping horse' display call is only heard on its breeding grounds in the far north of Europe.

Lymnocryptes minimus

UK STATUS Fairly common if little seen wintering visitor, when there may be 100,000 birds spread widely.

HABITAT The boggiest of grassland and saltmarsh, where it is expert at remaining hidden among the tussocks.

Snipe

Tucked away in the middle of boggy ground, this can be a difficult bird to see. Fortunately, in the breeding season, it has some of the most evocative bird sounds to seek out at dusk on moors and marshes, especially the incredible 'drumming' in display flight, plus it has a sharp-sneeze flight call should you accidentally flush one in winter.

STAR SOUND: ADVERTISING 'DRUMMING' Usually at nightfall and on moonlit nights, males fly high around the edge of their territory, diving steeply for a few seconds and then climbing again. During the dive, their flared outer tail feathers vibrate, creating a wonderful sound known as 'drumming', like a throbbing hum. The noise lasts around 1.5–2 seconds, increasing in volume and rising a little in pitch.

ADVERTISING *CHUK-A-CHUK-A* CALL Males sit either hidden in the marsh or sometimes on a post or mound, delivering a metronomic, staccato *chuk-a-chuk-a-chuk-a* call, the 'a' slightly higher in pitch, at about two cycles a second. This can go on for a minute or so. It is the third sound on Track 11.

TAKE-OFF CALL When flushed, it almost always 'sneezes' a couple of seconds after taking off, a retching *skatch!*

GROUND CALL Breeding birds sometimes give extended monotonous series of hard calls, *djeg djeg djeg djeg….*

Gallinago gallinago

UK STATUS Widespread breeding visitor, with about 80,000 pairs, although numbers have declined dramatically in the south and west. Very few drumming Snipe are left in southern England, with the New Forest (Hampshire) being the best of what remains there. Also a passage migrant and winter visitor from mainland and northern Europe, giving a winter population of probably more than 1 million birds.

HABITAT Breeds on wet moors, bogs, heaths and wet meadows; in winter, found in a wider range of wetlands, even small pools and ditches on occasion.

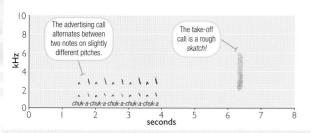

Red-necked Phalarope

This small, needle-billed wader has an unusual breeding strategy in which the duller males incubate the eggs and raise the young. Elsewhere, fewer than 50 are seen each year on passage at mainly coastal wetlands, with most seen in spring when in its largely grey breeding plumage with smart red, grey and white head and neck.

CALLS Typically short *chip* or *chik* notes, sometimes a rapid *chititiv*, the females' display calls being little more than an extended chatter of those notes.

Phalaropus lobatus

UK STATUS The small UK breeding population of around 50 pairs winters in the eastern Pacific, migrating across the Atlantic.

HABITAT Breeds on shallow mires.

WHERE TO HEAR Loch of Funzie, Fetlar (Shetland).

Grey Phalarope

Autumn gales force this scarce ocean-going wader close to shore as it migrates from the Arctic to southern oceans, with a few exhausted individuals dropping into coastal pools or even ending up inland on reservoirs, where they can be incredibly confiding, swimming with fast, jerky movements. Pale grey with black bandit masks, they are best told from Red-necked Phalaropes by their thicker bills.

CALL Occasionally, the flight call is heard in the UK, a sharp, metallic *vit*. The rather wheezy twittering display calls are reserved for their Arctic breeding grounds.

Phalaropus fulicarius

UK STATUS Mainly an autumn storm-driven migrant, in varying numbers but with a few hundred seen each year.

HABITAT Either flying along the coast or on shallow pools and lake edges.

Common Sandpiper

Along upland rivers and reservoirs in summer, or muddy pools in spring and autumn, the call is often the first sign of this small brown wader as it takes off and skims low over the water surface. Once landed, it pumps its rear end up and down nervously.

CALL The most typical call is a high *swee-swee-swee*, with often three to five notes but sometimes 10 or more, the series dropping slightly in pitch. Think of the Three Little Pigs who ran *swee swee swee*, all the way home! In alarm, it becomes a single drawn-out *sveeeeee* or *seeeek*.

DISPLAY CALL With the same high pitch and timbre as the call, the display call has a rapid pulsing, repeated cycle, rising and falling slightly in pitch, sounding like a squeaky bicycle wheel turning quickly: *svi-si-si-sit-sit svi-si-si-sit-sit*....

COMPARE WITH Green Sandpiper (call); Wood Sandpiper (call).

Actitis hypoleucos

UK STATUS Summer visitor, with around 15,000 breeding pairs mainly in western and northern Britain, and more widespread on migration. Fewer than 100 individuals winter in southern England.

HABITAT Breeds along upland rivers and northern loch and reservoir shores. On passage, visits freshwater wetlands with muddy margins. In winter, found on the lower reaches of tidal rivers.

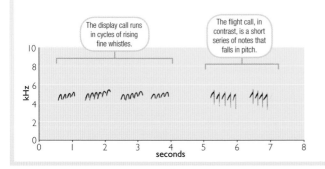

The display call runs in cycles of rising fine whistles.

The flight call, in contrast, is a short series of notes that falls in pitch.

Green Sandpiper

The call is an excellent clue to the identification of this rather solitary bird of muddy ditches and pools, especially when you accidentally flush one and it rises steeply into the air away from you.

FLIGHT CALL On passage, tends to call on take-off and sometimes in flight, with a consistently cheery or slightly panicky feel and a core *sveet* note. Variations include a simple *sveet*, a double-note *tsū-veet* or, most typically, *sveee sve-sveet*, with a long-short-long rhythm. The consistent thread is that all the notes tend to rise strongly in pitch.

ALARM CALL A straight-pitched *vit! vit! vit! vit!....*

Tringa ochropus

UK STATUS Fairly common passage migrant, in small numbers in spring and larger numbers (several thousand in total) in autumn. Fewer than 1,000 birds stay to winter in southern, lowland Britain. One or two pairs occasionally breed in northern Scotland.

HABITAT Uses a wide range of freshwater wetlands, including pool margins, shallow rivers, watercress beds and sewage farms.

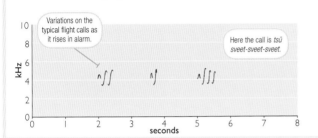

Variations on the typical flight calls as it rises in alarm.

Here the call is *tsū sveet-sveet-sveet*.

Wood Sandpiper

This dainty freshwater wader is an autumn treat with a flight call worth learning. The display call is only heard in northern Scotland.

FLIGHT CALL Most often heard on take-off, it gives three or four simple, repeated notes, with a sweet, baby-bird-like sound, *ship-ship-ship*. The 'p' at the end of each is soft, almost an 'f'. In alarm, the note is sharper and repeated at length.

'TO YOU' DISPLAY CALL Often given in fluttering, switchback flight, it delivers cycles of fast repeated, sweet couplets, *tulyū tulyū tulyū...*, overall lasting maybe 3–4 seconds, with some of the character of the Woodlark song but on a level pitch (page 210).

Tringa glareola

UK STATUS Passage migrant in small numbers in spring and especially autumn. About 25 pairs breed in northern Scotland.

HABITAT On passage, mainly freshwater pools with muddy margins; in the breeding season, it uses tree-fringed marshes and bogs.

Spotted Redshank

In breeding plumage, this is a stunning wader: velvety black with silvery dotted upperparts. However, most birds we see are plainer migrants or wintering birds, so knowing the main call can really aid identification.

FLIGHT CALL A rather consistent, bright *chew-it*, which is distinctive in its noticeable fast drop-and-rise pitch.

Tringa erythropus

UK STATUS Passage migrant in small numbers; fewer than 100 in winter.

HABITAT Mainly freshwater pools along the coast and estuary channels.

Redshank

A frequent sight on muddy estuaries and tidal creeks and breeding on wet grasslands, this is the 'sentinel of the marshes', prone to calling at the slightest provocation, with a wide repertoire of piping calls.

FLIGHT CALL This very typical call is a three-note $^{t}y\bar{u}$-$p\bar{u}$-$p\bar{u}$, the first note is sharper and a whipped downslur compared to the next two. It sometimes gives two-note or four-note versions, and also a longer, loud *tyoooo*.

ALARM CALL A sharply downslurred, clipped $^{t}yip!$, often delivered in long, uneven series when perched on a gatepost or mound. If flushed or panicked, it throws in a hysterical, fast $^{t}y\bar{u}h\bar{u}h\bar{u}h\bar{u}h\bar{u}$.

DISPLAY CALL Delivered in display flight or from the ground, the rolling repeated, rhythmic couplets have a lovely piping timbre, *t'yoo t'yoo*…, repeated for up to 20 seconds. Other versions include a rolling *WEE-lee WEE-lee*…, or *tū-LEE tū-LEE*… or a faster-rolling *tū-dlee tū-dlee*…. It is similar to the Greenshank display call but is more musical. On landing, it often concludes with a delightful, rolling *talūd'l talūd'l*….

COMPARE WITH Greenshank (song and call); Wood Sandpiper (song); Spotted Redshank (call).

Tringa totanus

UK STATUS Resident breeder, scattered widely but commonest in the north, with about 25,000 pairs; also a passage migrant and winter visitor, mainly from Iceland, giving a wintering population of about 130,000 birds.

HABITAT Breeds on coastal marshes, wet meadows, and northern moors and islands. Winters mainly around the coast on muddy estuaries.

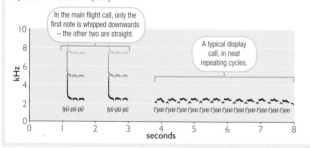

In the main flight call, only the first note is whipped downwards – the other two are straight.

A typical display call, in neat repeating cycles.

Little Gull

This scarce and petite gull delicately picks tiny food items off the surface of the water in dipping, buoyant flight. Adults have dark grey underwings; young birds have a bold black 'W' across their wings. The prime site is Hornsea Mere (East Yorkshire), where up to several thousand European birds gather to moult in late summer, but in 2016 a pair at Loch of Strathbeg (Aberdeenshire) was the first to successfully fledge young in the UK.

CALLS Only occasionally heard in the UK, the most typical calls are a short, often repeated *k' k' k' kek!*, sometimes sounding Jackdaw-like, or a more toy-trumpeting *chow!* or even *pow!*

Hydrocoloeus minutus

UK STATUS Scarce passage visitor in spring and from late summer onwards, and a few stay in winter.

HABITAT Mainly coastal, with a few turning up on inland waters.

Greenshank

This pale and rather robust wader has a triple-note flight call, while its cyclical display call is one of the wild sounds of the northern bog country.

FLIGHT CALL Most often when taking off or landing, and in flight, gives a strong, confident, $^{ty}\bar{u}$ $^{ty}\bar{u}$ $^{ty}\bar{u}$, audible at distance. It is very similar to the $^{ty}\bar{u}$-pū-pū of the Redshank, but each syllable is identical. Sometimes it will only give one or two notes, or run them into a longer series.

ALARM CALL A rather clipped *tyup!*, about the same pitch as the flight call but harder, sometimes in a long fast series, *tyip-tyip-tyip-tyip-tyip…*.

DISPLAY CALL Usually in flight, it gives repeated couplets, $ch\bar{u}$-wee $ch\bar{u}$-wee.., two to three per second, each couplet dipping in pitch in the middle. Often there is a momentary hiccup in between: $p'ch\bar{u}$-ee $p'ch\bar{u}$-ee… or $c'l\bar{u}$-ee $c'l\bar{u}$-ee…, giving it a wonderfully rhythmic feel.

COMPARE WITH Redshank (call and song).

Tringa nebularia

UK STATUS Summer visitor to northern Scotland where around 1,000 pairs breed. Passage migrant in spring and autumn elsewhere, and about 700 winter thinly along south and west coastal areas.

HABITAT Breeds on wild blanket bogs with lots of pools and streams. On passage, visits muddy marshland pools and coastal wetlands. In winter, found mainly on estuaries.

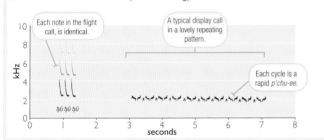

Each note in the flight call, is identical.

A typical display call in a lovely repeating pattern.

Each cycle is a rapid *p'chu-ee*.

tyū tyū tyū

kHz — 10, 8, 6, 4, 2, 0

seconds — 0, 1, 2, 3, 4, 5, 6, 7, 8

Black-headed Gull

The hoarse and insistent calls of this familiar and common gull are an abiding sound of marshes, estuaries and even school playgrounds in winter, while its dense breeding colonies are cacophonous.

LONG CALL This often heard and distinctive call is a drawn-out, descending *RAAAARRRRGH*, or sometimes more *ORRRRRRGH* or *REEE-ARRRRGH*. It is harsh, grating and loud, and may include gargling sounds. It is used in aggression and to signal territory.

SHORT CALLS It often intersperses the long call with short r*arr rarr rarr…* calls or an even shorter, barked *row! row!*, all as hoarse and rasping as the long call. It also often makes short *kek* or *kek-kek* calls, especially when squabbling.

COMPARE WITH Mediterranean Gull.

Chroicocephalus ridibundus

UK STATUS Resident breeder, with around 140,000 pairs in colonies. Widespread in winter, with many birds arriving from Europe and more than 2 million birds in total.

HABITAT Breeds in a wide range of wetlands including bogs, coastal marshes and gravel pits. Otherwise, ranges across lowland areas, coastal and inland, roosting on lakes and reservoirs and loafing on playing fields.

Kittiwake

This seafaring gull with black wing tips says its own name, one of the abiding sounds from massed, cliff-face seabird colonies.

STAR SOUND: 'KITTIWAKE' CALL When birds fly too close to a neighbour's nest or try to take over another bird's ledge, there is a flurry of increasingly excited and full-volume *kittiWAYK kittiWAYK kittiWAYK* calls. At its most intense, the ending is a shrill yodel, *kittiWAY-y^{uk}*. It takes a while for things to calm down again!

OTHER CALLS Despite the apparent ruckus, many birds in a colony actually sit silently for long periods. Sometimes they make little high yelping *kap* or *kop* calls, like toy trumpet toots, or even a long, wavering *arrgghhh*. There are also moments of total silence during a 'dread', when every bird falls instantly quiet and flies out from the cliff face.

Rissa tridactyla

UK STATUS Summer visitor to coastal cliffs, with a population of around 400,000 pairs. Numbers have fallen dramatically at many sites.

HABITAT Vertical coastal cliffs, requiring plentiful ledges and safety from ground predators. Otherwise marine.

WHERE TO HEAR Bridlington town centre (East Yorkshire); Newcastle quayside (Tyne and Wear), including the bespoke Kittiwake Tower in Gateshead; Seaford Head (East Sussex); Mumbles Pier (Swansea); Bempton Cliffs (East Yorkshire); Farne Islands (Northumberland); and Rathlin Island (Ulster), plus many seabird colonies around Scotland.

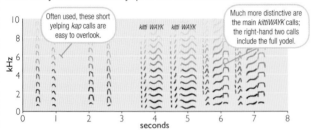

Often used, these short yelping *kap* calls are easy to overlook.

kitti WAYK   kitti WAYK

Much more distinctive are the main *kittiWAYK* calls; the right-hand two calls include the full yodel.

Common Gull

It is only in Scotland that the Common Gull is anywhere near 'common' as a breeding bird. Although very vocal in the breeding season, it is rather quiet away from its colonies.

SHORT CALLS The scientific name means 'dog gull', based on its short, thin, sharp *r^{o}w!* barks, like a tiny (but nevertheless loud) dog. Also makes harder *ruk!* noises.

LONG CALL A repreated pure, ultra-shrill *re^{-eeee}-er*, leaping ear-splittingly at the start. Pairs sometimes give this call in duet at the nest site.

Larus canus

UK STATUS Just under 50,000 breeding pairs, mainly in Scotland, spreading out more widely in winter and joined by many birds from northern Europe, with a wintering population of around 700,000 birds.

HABITAT Breeds in colonies on wet moors and bogs and northern islands, often with other gull species. In winter, mainly coastal but also on downland fields.

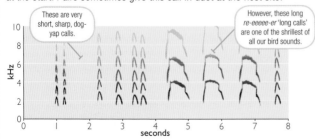

These are very short, sharp, dog-yap calls.

However, these long *re-eeeee-er* 'long calls' are one of the shrillest of all our bird sounds.

Mediterranean Gull

A relatively new arrival, it now nests quite widely and some southern colonies contain several hundred pairs. The adults' pure white wing tips stand out in flight, while in breeding plumage their hoods are truly black, their bills bright red and stout, but it is their far-carrying 'cat calls' that are often the first thing to draw your attention.

'CAT CALL' This very diagnostic call is a simple, cute, mewing $y^{o}w$, rising steeply in pitch and falling again. If excited, the call may start with shorter notes, such as *wha wha* or *yō*. It is vocal at the nest site and often calls in flight in spring and summer but is much quieter in winter.

Ichthyaetus melanocephalus

UK STATUS Rapidly increasing resident, with more than 1,000 pairs, most in coastal southern and eastern England. Also a passage migrant through the English Channel.

HABITAT Usually breeds among Black-headed Gulls on shingle islands and saltmarsh, feeding inland on worms in fields. Winters around estuaries.

WHERE TO HEAR Poole Harbour (Dorset); Langstone Harbour (Hampshire); Pagham Harbour (West Sussex); Rye Harbour (East Sussex); Medway Estuary (Kent); Minsmere (Suffolk).

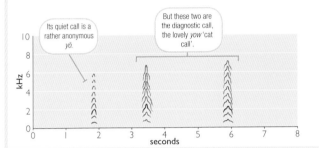

Its quiet call is a rather anonymous *yō*.

But these two are the diagnostic call, the lovely *yow* 'cat call'.

Great Black-backed Gull

This bulky bird has essentially the same repertoire of calls as the Herring and the Lesser Black-backed but with the might and gruffness you'd expect from a heavyweight, and is often much less vocal.

LONG CALL This display call starts with long build-up notes that break into a series of hoarse laughing notes, as braying as a donkey. It gives these notes with its beak pointing up at the sky.

SHORT CALL A deep, gravelly or throaty $c\bar{o}www$ or $^{rr}\bar{o}www$, with downwards inflection.

ANXIETY CALL *Bo-bo-bob*, as the Herring Gull but deeper.

COMPARE WITH Herring Gull; Lesser Black-backed Gull.

Larus marinus

UK STATUS Largely sedentary resident, with about 17,000 pairs around our northern and western seaboard. Northern populations disperse southwards in winter.

HABITAT Mainly breeds on coastal islands and headlands, and most feed on the coast although small numbers wander inland in winter.

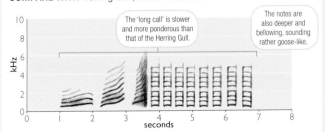

The 'long call' is slower and more ponderous than that of the Herring Gull.

The notes are also deeper and bellowing, sounding rather goose-like.

Herring Gull

The sound of the seaside, especially with its triumphant display laugh, this is an extremely vocal bird with a large repertoire of calls, which vary hugely in intensity and volume as its mood changes.

LONG DISPLAY CALL A drawn-out series of wild, triumphant laughing shrieks. It starts with a couple of low-pitched notes, its beak pointing at the ground, and then throws its head back and breaks into two or three long, shrill notes, followed by a run of faster laughs that finally slow and calm.

ANXIETY CALL Near the nest, gives a nervous *bo-bo-bob*.

SHORT CALL Typically *k'yow*, much repeated and sharper in alarm.

LONG 'MEW' CALL Each note is a long, whining, loud *aaaarrrrrgghhh*.

Larus argentatus

UK STATUS Widespread but overall declining resident, numbers having halved since the 1960s to about 130,000 breeding pairs, although increasing in urban areas. Population boosted in winter by birds from mainly Norway.

HABITAT Predominantly coastal but increasingly inland in the breeding season, nesting on rocky cliffs, shingle beaches, gravel pit islands and rooftops. In winter, widespread on agricultural land and refuse tips, roosting on reservoirs or the sea.

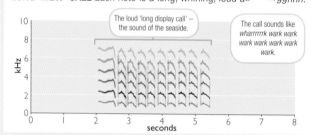

The loud 'long display call' – the sound of the seaside.

The call sounds like *wharrrrrk wark wark wark wark wark wark wark wark*.

Lesser Black-backed Gull

Slightly smaller than the Herring Gull, adults have dark grey upperparts and yellow legs. It has a similar range of calls to the Herring, but they are typically faster, less shrieking and more nasal and muffled, as if they are holding their 'hands over their faces'. They are very vocal at their nesting colonies but rather quiet otherwise.

LONG CALL Similar in pattern to the long call of the Herring Gull, the laugh is typically faster, more nasal, and usually shorter in overall length.

SHORT CALL Rather nasal *owp*, *owwwh* or *chow*, rather like the Common.

ANXIETY CALL *Chap-chap-chap* or *wap-wap-wap*.

COMPARE WITH Herring Gull; Great Black-backed Gull.

Larus fuscus

UK STATUS About 110,000 pairs breed, with many just dispersing short distances in winter but others moving to Iberian and African coasts.

HABITAT Once just a coastal breeder, it has now colonised some rooftops inland. Widespread in winter, often on landfill sites and farmland by day, roosting on reservoirs.

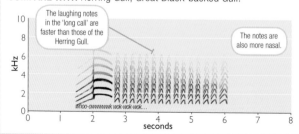

The laughing notes in the 'long call' are faster than those of the Herring Gull.

The notes are also more nasal.

whoo-owwwwwk wok-wok-wok...

Sandwich Tern

This is the largest of our terns with a black shaggy cap and yellow-tipped black bill. It alerts you to its presence with its loud flight calls, and its colonies are a mass of noise.

KERRIK **CALL** The typical call is a very harsh, loud *KERRIK*, which is either straight in pitch, or can rise or fall on the second syllable. Some have a particularly shrill ending, *KERRIX*. It is a sound in which you can simultaneously hear shrill high components and deeper grating sounds. It also gives more conversational, short *iks* and *eks*, but can rarely stop itself from breaking into the full call.

RASP CALL Sometimes gives a longer, simpler rasp, *errrrrrrk*.

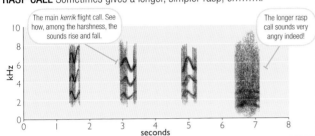

The main *kerrik* flight call. See how, among the harshness, the sounds rise and fall.

The longer rasp call sounds very angry indeed!

Thalasseus sandvicensis

UK STATUS Summer visitor around the coast, with only about 25 colonies containing a total of 12,000 pairs. Just a handful remain here in winter.

HABITAT Breeds in densely packed colonies on low, flat islands and coastal lagoons. Rarely seen inland.

WHERE TO HEAR Brownsea Island (Dorset); Pagham Harbour (West Sussex); Rye Harbour (East Sussex); Blakeney Point (Norfolk); Cemlyn Lagoon (Gwynedd); Hodbarrow (Cumbria); Farne Islands (Northumberland); Isle of May (Fife); Lower Lough Erne (Co. Fermanagh).

Little Tern

A Little Tern looks minute compared with other terns and gulls, barely larger than a Starling, although with longer, incredibly slender wings. In flight, it has ultra-fast flickering wingbeats, often hovering expertly before plunge-diving for fish. The Swallow-like calls are distinctive.

DISPLAY CALL Vocal in and around the nesting colony, the display call is a rapid (as if speeded up), rather sweet and long chittering in random bursts, often given by a male as he presents a fish to his mate. It is high-pitched and with some rhythmic elements, along the lines of a very fast *chididit-wijit-wijit-weee-teee....*

CALL A high *w$^{ik!}$, whee^k* or *wijit*, often in alarm.

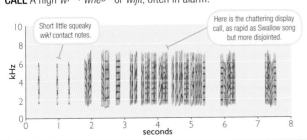

Short little squeaky *wik!* contact notes.

Here is the chattering display call, as rapid as Swallow song but more disjointed.

Sternula albifrons

UK STATUS Summer visitor from West Africa to around 60 scattered colonies, with about 1,500 breeding pairs. Very rare inland.

HABITAT Breeds on shingle islands and sandy beaches, often close to the high tide line, preferring areas free of vegetation. Feeds in inshore waters.

WHERE TO HEAR The Fleet (Dorset); Eccles and Winterton (Norfolk); Kessingland and Benacre (Suffolk); Pagham Harbour (West Sussex); Rye Harbour (East Sussex).

Common Tern

The trio of Common, Arctic and the rare Roseate Tern are all slender, pale terns, buoyant in flight, with black caps, red legs and long tail streamers. All are noisy at their colonies, where the three species can breed alongside each other, and where separating out individual noises from a colony's cacophony can be difficult. Listen too for 'dreads', when a whole colony of terns, sensing danger, falls instantly silent and rises into the air. All three species also call regularly in flight, although less so on migration.

The Common is the only one you'll see in summer over southern lakes.

ALARM CALL Sometimes drawn out, sometimes faster and repeated, it is a two-syllable *eeer-yarrr*, dropping to a lower pitch on the second syllable, with a heavy rasp throughout; alternatively, the slur can be upwards, as in *eer-yurr eer-yurr*.

IK-IK CALL Often repeated in well-spaced, uneven series, this very sharp, snapped *ik! ik! ik! ik! ik!...* can be inserted between other calls.

ANGER CALL A deeper, very fast *kek'ek'ek'ek'ek*, especially when mobbing predators.

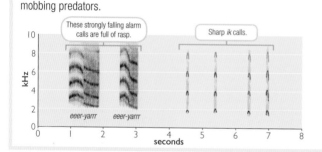

These strongly falling alarm calls are full of rasp.

Sharp *ik* calls.

eeer-yarrr *eeer-yarrr*

Sterna hirundo

UK STATUS Summer visitor spread thinly across central and eastern England and around northern coasts and Irish loughs, with around 12,000 breeding pairs.

HABITAT Mainly coastal islands in northern England and Scotland, often among Arctic Terns, but regular at freshwater reservoirs, gravel pits and lakes in central and eastern England.

WHERE TO HEAR Great views can be had at colonies such as: Brownsea Island (Dorset); Rye Harbour (East Sussex); Rye Meads (Hertfordshire); Rutland Water (Rutland); and Portmore Lough (Co. Antrim).

Roseate Tern

The best way to see this rare, pale, pink-flushed tern is to take a boat trip out from Amble (Northumberland) to Coquet Island, where about 100 pairs breed. However, landing is not allowed, and as they nest among screaming hordes of Arctic and Common Terns you'll need to concentrate to hear their calls.

ALARM CALL The most distinctive call is a deep *krrraa*, straighter pitched than the familiar *eer-yarr* of the Common or *peeeearr* of the Arctic, at times incredibly like a Jay.

ANGER CALL Very typically a double-note *bho-ik* or *chirrik*, reminiscent of the Sandwich Tern's *kerrik* but higher pitched.

COMPARE WITH Arctic Tern; Common Tern.

Sterna dougallii

UK STATUS Rare summer visitor, with just three main breeding colonies in the British Isles: two in Ireland and one on Coquet Island.

HABITAT Marine, coming ashore to breed and occasionally rest on beaches.

Arctic Tern

Anyone who has visited the Arctic Tern colony on Inner Farne
(Northumberland) will know only too well the experience of being
dive-bombed by the brave parent birds, one of the most exciting (or
unnerving!) and loudest bird spectacles in Britain. This species is also
rightly famed for its vast migrations, wintering near the Antarctic.
Compared with the Common and the Roseate Tern, it has a blood-red
bill and greyer underparts. It has a very similar suite of calls to the
Common Tern, and as variable, many being higher pitched, but the
differences are slight and need experience to be used with confidence.

ALARM CALL A buzzy *peearrrrr*, starting higher and tending to drop
further in pitch than the *eeer-yarrr* call of the Common.

CHIK CALL More a *chik! chik!*... or *chek! chek!*... than the
Common's *ik! ik!*....

ANGER RATTLE A very fast *tek-ek-ek-ek-ek-ek* with a sharp, clicking
quality, rather like a Geiger counter.

KITTA-KERRR CALL Often makes a rhythmic *kitta-kerrr kitta-kerrr*
when returning to its nest.

Sterna paradisaea

UK STATUS Summer visitor, with just
over 50,000 pairs, the vast majority
from Northumberland northwards;
passage birds are mainly seen in
spring, a few passing overland but most
moving through quickly offshore.

HABITAT Breeding colonies are mainly
on small islands and remote beaches;
otherwise marine.

WHERE TO HEAR Inner Farne and
Long Nanny (Northumberland); Isle of
May (Fife); Outer Hebrides; Orkney;
Shetland.

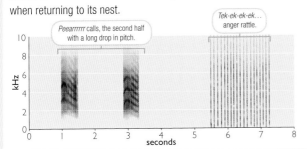

Peearrrrr calls, the second half with a long drop in pitch.

Tek-ek-ek-ek... anger rattle.

Black Tern

Such an airy and buoyant tern, it flits lightly over the water, as if
on marionette strings, dipping down to the water surface to snatch
small food items. The adults in spring are very distinctive with their
charcoal-coloured heads and bodies, while in autumn the adults and
juveniles have dark caps but white bodies. They don't breed in the UK,
so it takes a trip to the continent to see them at their wetland nesting
colonies where they are at their most vocal.

CALL Very occasionally in the UK they will give flight calls, which are
harsh, rather high-pitched, short, sharp *chit* and *churt* sounds.

Chlidonias niger

UK STATUS Passage migrant in both
spring and autumn, mainly in southern
parts.

HABITAT Stops off at reservoirs and
lakes, and good numbers also pass up
the English Channel in spring.

Great Skua

A trip to one of the UK's few Great Skua colonies is nerve-wracking: accidentally stray too close and this hefty bird will fly at you in a low-level attack with speed and menace, and is not afraid to whack you with its hooked bill should you not take evasive action. Only here on the breeding grounds are you likely to hear them; otherwise, they are largely silent and most views are of them way offshore, looking like an all-brown chunky gull with bold white wing flashes, as they mercilessly harry other seabirds to force them to regurgitate food.

DISPLAY CALL Stood near the nest, adults make a repeated nasal *snarp! snarp!*, about two to three per second, while holding their wings up vertically above their backs in a dramatic posture.

OTHER CALLS If agitated at the nest, adults make ominous *bo-bo-bok* calls. Full attack is often in ominous silence, except for a great rush of wings as it skims your head!

Stercorarius skua

UK STATUS Summer visitor to the far north, with around 10,000 breeding pairs. Otherwise, a passage migrant in spring and especially autumn offshore, rarely seen over land.

HABITAT Breeds on coastal moors overlooking the sea; otherwise exclusively marine.

WHERE TO HEAR Handa (Sutherland); Hoy (Orkney); Fair Isle and Hermaness (Shetland).

Arctic Skua

Most views of the Arctic Skua are of a dashing dark seabird with long, pointed wings, twisting and turning in dogged pursuit of a luckless tern or other seabird, forcing its quarry to disgorge its catch of fish. It has two main plumage types, either dark grey-brown all over ('dark phase') or with pale belly and neck ('pale phase'). However, it is only on their few breeding grounds in the far north that you are likely to hear them, for they are otherwise largely silent.

DISPLAY CALL Fairly vocal on the breeding grounds; the call is remarkably Kittiwake-like, an *ee-^{y}ow*, which often yodels into the *-oww*.

ALARM CALL A sharp, *y^{o}w! y^{o}w! …*, or *y^{o}w-o^{w}-ow*, rather like the Common Gull in feel.

OTHER CALLS Sometimes makes a long series of short *yak yak yak…* calls, and when aggressively swoop-bombing predators near the nest (including humans!) it will make sharp *chowp!* calls.

COMPARE WITH Kittiwake; Common Gull.

Stercorarius parasiticus

UK STATUS Declining breeding visitor to northern Scotland, with around 2,000 pairs. Also regularly seen as a passage migrant offshore, especially down the east coast of Scotland and England, lingering more in autumn.

HABITAT Breeds on open moors usually close to the sea; otherwise exclusively marine.

WHERE TO HEAR Handa (Sutherland); Orkney.

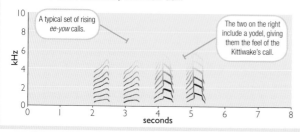

A typical set of rising *ee-yow* calls.

The two on the right include a yodel, giving them the feel of the Kittiwake's call.

Guillemot

Chocolate-brown and fine-billed in comparison to Razorbills, Guillemots are silent much of the year when at sea. However, for a brief few months in spring and summer, they are one of the standout sounds at our seabird colonies, making a monkey-house range of sounds not to be missed.

STAR SOUND: CALLS Packed onto their breeding ledges, the birds sound like a theatre audience having a good belly laugh. Surging calls include repeated chuckles, *ar ar ar ar ar...*, and lower, monkey-like noises, *ooh ooh ooh...*. However, most obvious are bold, loud *rr^{a}ar* calls and a very noticeable *wrr^{a}rr^{r}r^{r}r^{r}r*, which ends in a wild gargle, usually given when an adult attempts to land in the wrong place among territorial birds. By summer, the calls are interspersed with high-pitched begging calls from the chicks.

COMPARE WITH Razorbill; Puffin.

Uria aalge

UK STATUS Summer visitor to northern and western coasts, especially rat-free islands, with just under 1 million pairs; disperses in winter to seas around northwestern Europe.

HABITAT Usually nests on more exposed, sheer, perilous cliff ledges than Razorbill; otherwise exclusively marine.

WHERE TO HEAR As for Razorbill.

Razorbill

Razorbills and Guillemots are like the flying penguins of the north, dark above and white below, awkwardly upright on land but in their element underwater. Razorbills are blacker above than Guillemots, and with a more substantial bill dissected by a vertical white groove, although separating the two can be difficult at a distance. In contrast to noisy Guillemots, Razorbills are much less obvious vocally, but their deep growls are well worth listening for.

CALL The limited repertoire is largely low-pitched, their slow, creaking, ratchety growls easy to miss. Compared to the groans of the Puffin, the croak is drier and more wooden, the individual notes not merging together. The calls are given in greeting and in bonding by pairs at their nests, and in spring small groups gather on the water beneath the nesting cliffs, tails cocked, where they growl enthusiastically.

COMPARE WITH Guillemot; Puffin.

Alca torda

UK STATUS Summer visitor to northern and western islands and sea cliffs, with around 130,000 pairs; disperses in winter around the seas of northwestern Europe, often within sight of land.

HABITAT In the breeding season, nests on rather hidden and sheltered cliff ledges; otherwise exclusively marine.

WHERE TO HEAR Farne Islands (Northumberland); Treshnish Isles (Argyll and Bute); Skomer and Skokholm (Dyfed); Rathlin Island (Co. Antrim).

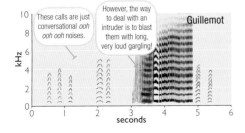

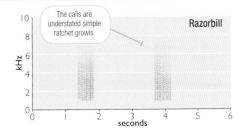

Black Guillemot

Known in Shetland as the Tystie, this velvety black auk has large white wing patches in summer, while in winter it is a hotchpotch of pied markings. However, in all seasons it has the brightest red feet. It nests in small, loose colonies, and hearing its curious high-pitched calls can be difficult except in a few northern harbours.

CALLS It can be vocal at its breeding sites, but its ultra high-pitched, thin whistles, very different to those of other auks, are easily overlooked. They include: a long, drawn-out, straight and very high-pitched *sweeeeeeeen*, like the ringing sound when running your finger around the rim of a crystal glass; a series of fast, short notes, *swip swip swip swip swip…*, like a high-pitched Common Sandpiper's call, sometimes so fast as to be a twitter, while exaggeratedly head-nodding; and a series of fast seesawing high notes given in a penguin-strutting display *see-ʷᵉ-see-ʷᵉ-see-ʷᵉ….*

COMPARE WITH Rock Pipit; Common Sandpiper.

Cepphus grylle

UK STATUS Largely sedentary, with fewer than 20,000 pairs, most around north and west Scottish and Irish coasts.

HABITAT Marine, breeding among cliff boulders or in holes in piers and harbour walls.

WHERE TO HEAR Peel Harbour (Isle of Man); Portpatrick Harbour (Dumfries and Galloway); Oban Harbour (Argyll and Bute); Orkney; Shetland.

Puffin

With their clown-like expression and upright posture, Puffins are a crowd-pleasing favourite, and a trip to an island colony is unforgettable. They are not very vocal birds, even at their nesting sites, but up close the groaning calls are as entertaining as their looks.

CALL When mates greet, often inside their burrows, they give an amusing if quiet deep, drawn-out creaking groan, often several seconds long, like the sound of a distant chainsaw, which often rises and then falls in pitch and volume. It can seem to rumble through the ground. Sometimes, it becomes exaggerated into a ponderous and most entertaining guffaw, descending the scale, *hawww hawww hawww….* Pairs also enthusiastically rattle their bills together in greeting.

COMPARE WITH Razorbill; Guillemot.

Fratercula arctica

UK STATUS Summer visitor, April–July, with almost 600,000 pairs, mainly in the north and west, although sadly numbers are dropping and it is now a Red-listed species of conservation concern. Winters far out in the Atlantic.

HABITAT Nests in burrows that they dig in soft soil on clifftops and steep grassy banks on rat-free islands with direct access to the sea; otherwise exclusively marine.

WHERE TO HEAR Skomer (Dyfed); Farne Islands (Northumberland); Treshnish Isles (Argyll and Bute); Fair Isle (Shetland); Rathlin Island (Co. Antrim).

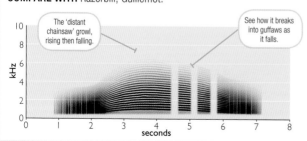

The 'distant chainsaw' growl, rising then falling.

See how it breaks into guffaws as it falls.

Feral Pigeon (Rock Dove)

Track 32

The 'town pigeon' and its wild ancestor, the Rock Dove, have identical calls. See page 74 for full description.

SONG The male repeats *kukker K'ROO*, the first notes staccato, the ending a muffled bubbling.

NEST CALL A simple, deep, groaning *OOOOO-uh* in series, with a long first syllable.

WING CLAPPING As a male flies out from his nest site, he claps his wings several times.

COMPARE WITH Stock Dove.

Columba livea

UK STATUS Sedentary resident, with about 500,000 pairs of Feral Pigeon, but wild Rock Doves are only found in north and west Scotland and Ireland.

HABITAT Urban centres, farmyards, arable crops and cliffs.

Stock Dove

Track 71

This smart and compact small pigeon has a soft, pumping song. See page 106 for full description.

SONG A series of 4–12 deep-pitched, gentle *whoo-er-wup* notes.

WING CLAPPING In display flights a male claps his wings over his back rather weakly two to three times.

COMPARE WITH Feral Pigeon.

Columba oenas

UK STATUS Sedentary resident in lowland areas and absent from the far north and west, with more than 250,000 breeding pairs.

HABITAT Farmland and country parkland, needing ancient rotting trees or quiet barns for nesting.

Woodpigeon

Track 30

This is a staple songster of woods and gardens everywhere. See page 72 for full description.

SONG The deep, five-note cooing phrase, best remembered as 'I DON'T want to go', is repeated three to four times.

CLOSE DISPLAY CALL A much quieter *mowwwwww mu-mu-mu mawwww*.

NEST CALL A quiet, groaning *mer-mawwww*.

WING NOISES In display flight, the male rises steeply, claps his wings once or twice, and glides down. Woodpigeon wings also clatter on take-off, and squabbling males slap each other noisily.

COMPARE WITH Collared Dove; Stock Dove; Feral Pigeon.

Columba palumbus

UK STATUS Mainly a sedentary and increasing resident with 5 million pairs; rare only in treeless upland areas.

HABITAT A cosmopolitan bird of farmland, woodland and, increasingly, urban parks and gardens, nesting in trees and feeding in fields and gardens.

Turtle Dove

It is increasingly hard to find our smallest dove, whose UK population has plummeted by around 98 per cent in 50 years. Its gentle song, after which it was named, was once a sound of summer in many southern areas.

STAR SOUND: SONG Males sing from trees and hedges, a lovely, warm, deep purring series of double notes, *trrrr trrrr*, each about half a second long, and reminiscent of 1970s telephone ringtones. Each male's song has subtly different rhythms and patterns. For example, the first note might be slightly longer and rising in pitch, the next three straight, giving *trʳʳʳʳʳ trrrr* (pause) *trrrr trrrr* (pause).

DISPLAY CALL When displaying directly to a female, males double the pace of the song.

Streptopelia turtur

UK STATUS Summer visitor, returning in May, with the majority now found in East Anglia, Sussex and Kent, with maybe only 10,000 pairs left. The calamitous declines are due to a lack of arable weed seeds and problems on their migration routes and wintering grounds.

HABITAT Arable and mixed farmland with mature hedgerows and weedy crops.

Collared Dove

Track 31

The lethargic song of this sandy-coloured dove is heard in gardens everywhere. See page 73 for full description.

SONG The deep-pitched *ku ᴷᴼᴼᴼ kū* is repeated several times to the rhythm of a bored football fan chanting 'u-ni-ted'.

'AIR' LANDING CALL A simple, breathy, long *aⁱʳʳʳʳʳʳʳʳ*.

WING NOISES The wings make a fine, pulsing noise in flight, and birds also wing-clap lightly on take-off and in flight display.

COMPARE WITH Cuckoo; Woodpigeon.

Streptopelia decaocto

UK STATUS Sedentary resident, with around 1 million pairs in mainly lowland areas, with signs of a small drop in the population in recent years.

HABITAT Largely associated with human habitation and farms, and commonest where there are abundant sources of grain such as farmyards.

Cuckoo

Track 87

The simple song of the male is the essence of spring. See page 120 for full description. However, the female has a beautiful call too.

STAR SOUND: SONG Males repeat *ku kū* many times in series. By summer, it can be a more hurried *ku-ku kū* or even an off-key, rising *ku k'wō*.

GOWKING CALL In high excitement or when mobbed, it makes a strange, quiet, throat-clearing call, a fast *chuk-chok-chak-chowk* known as 'gowking'.

BUBBLING CALL (FEMALE) Females sometimes bubble in flight, about 15–20 beautiful notes rising in pitch and volume before fading away.

COMPARE WITH Collared Dove (song); Green Woodpecker (call).

Cuculus canorus

UK STATUS Summer visitor from Africa, arriving from late April and early May, with most gone by high summer. The population is around 15,000 'pairs', widespread but thinly spread and declining.

HABITAT Mainly found wherever its host species live, so visit reedbeds for Reed Warblers, upland valleys for Meadow Pipits and heathland and hedgerows for Dunnocks. Rare in urban areas.

Barn Owl

Often emerging in the late afternoon, this ghostly white owl is sometimes known as the 'Screech Owl'. Your best chance of hearing one is near a nesting site at dusk early in the breeding season.

SCREECH CALL Lasting a second or more, it is typically a straight blast of noise, *shreeeeeeeeeee*, sometimes like a hiss, at others a dry Darth Vader rasp, or even like the scream of a creature in distress. The screech of territorial males typically has a vibrating quality.

OTHER CALLS Gives a wide range of intergrading calls, mostly in and around the nesting site, which vary from gentle 'snoring' calls from females to twittering from very young birds.

Tyto alba

UK STATUS Sedentary resident across much of lowland Britain, with a breeding population of around 4,000 pairs. Numbers are pegged back in harsh winters or when prey is in short supply.

HABITAT Mainly a bird of farmland, especially with plenty of hedgerows, 'beetle banks', rough grassy areas and wet grassland. Benefits from the provision of special nestboxes.

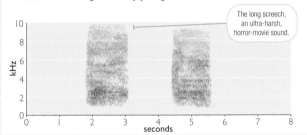

The long screech, an ultra-harsh, horror-movie sound.

Little Owl

With its cute expression and cuddly-toy shape, this is a favourite of many, often emerging in the late afternoon, and with two key calls.

CONTACT CALL A short, fairly high-pitched $h^{we}e\bar{o}!$, with a fast rise-and-fall, making it feel sharp and snatched. Sometimes it is given four to six times in series.

ADVERTISING TOOT From about March–June, males give a rather fun and simple $wooo^{o(k)!}$, rising in pitch and with a clipped ending, somewhat like the $kooo^{or}$ note of the Curlew.

ALARM CALL A fast, irregular, *chik-chik!   chik! chik-chik!....*

COMPARE WITH Curlew; Tawny Owl.

Athene noctua

UK STATUS Sedentary introduced resident, with around 5,000 pairs, most in lowland England. There have been considerable declines in many parts in recent years.

HABITAT Mainly mixed, low-intensity farmland in rural areas, including orchards and parkland, with plenty of mature trees and hedgerows, nesting in holes in trees, barns, stone walls and quarries.

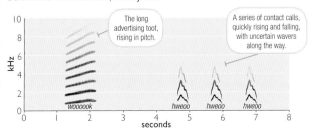

The long advertising toot, rising in pitch.

A series of contact calls, quickly rising and falling, with uncertain wavers along the way.

woooook   hweoo   hweoo   hweoo

Tawny Owl

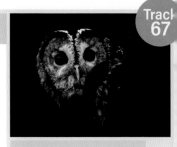

More often heard than seen, this common yet nocturnal owl has a famous contact call and hoot that, together, are probably the origin of the term 't'wit t'woo'. See page 102 for full description.

STAR SOUND: ADVERTISING HOOT From autumn until spring, males give deep melancholy hoots with three elements: a simple, long hoot (long pause), short, quiet note (short pause), finishing with a very long, quivering hoot. Females sometimes give a screechy, hoarser version.

CONTACT CALL *ke-WIK!*, short and rising very sharply in pitch, given by both sexes.

SNIPE-DRUM CALL Beautiful but little heard; males give a rapid series of *wah* notes, about 12 per second for 2–3 seconds, that grows in volume and then subsides, rather like the drumming call of the Snipe.

BEGGING CALL Chicks give insistent, lisping squeaks, like sucking in air through closed teeth, *sveet sveet sveet….*

Strix aluco

UK STATUS Sedentary resident with around 50,000 pairs found across almost all of mainland England, Scotland and Wales; absent in Ireland and other islands.

HABITAT A woodland bird, but also found in small copses, parks and large gardens.

Long-eared Owl

Our most elusive owl is rarely seen except at a few winter roosts, and few people get to hear the soft, deep nighttime hooting of the males. Listen, though, for the high, squeaky calls of chicks in summer.

ADVERTISING HOOT At night in late autumn and especially early spring, high in a tree, males give a deep, straight *hoo* just once every 2.5–4 seconds. It is as regular as a distant foghorn, the sound not carrying far. He may sometimes duet with his mate who gives a more kazoo-like and repeated *zoo* note, again widely spaced.

BEGGING CALL Older chicks make simple, very high-pitched loud squeaks, *seeee*, dropping strongly in pitch and often repeated, like a rusty gate swinging sharply on its hinges.

WING CLAPPING Males flying over their territory at night make single weak wing-claps.

Asio otus

UK STATUS Sedentary resident, widespread but thinly scattered, with an estimated 1,800–6,000 pairs, plus extra wintering birds from northern Europe.

HABITAT Breeds in copses and small plantations near open moors, heaths, downland and rough grassland. In winter, roosts by day in dense thickets, often in wetlands.

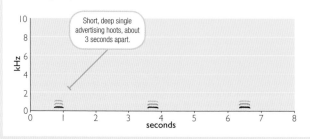

Short, deep single advertising hoots, about 3 seconds apart.

Short-eared Owl

Most often seen on a winter's afternoon quietly quartering over rough coastal grassland; you will need to visit its northern and upland breeding areas to have a chance of hearing the display call and wing clapping.

ADVERTISING CALL Only occasionally vocal, even in the breeding season; males sometimes give a rather fast series of 12–20 deep hoots, at a rate of about four to five a second: *oop oop oop*..., the series rising slightly in pitch and volume. It is most often given in display flight, circling high into the sky.

WING CLAPPING Also in display flight, males sometimes clap their wings beneath their body about 4–10 times very quickly.

BARKING CALL Both sexes can give a range of rasping barks, typically starting slow and ending sharply: *rrrrrap!* or *tcheeup!* In alarm, it can be given in faster series, *rraf-rraf-rraf-rraf.*

COMPARE WITH Long-eared Owl.

Asio flammeus

UK STATUS Breeds mainly in northern Britain, numbers varying but between 600 and 2,000 pairs. More widespread on passage and in winter, with extra birds arriving in very variable numbers from northern Europe.

HABITAT Breeds mainly on moors and in newly planted forestry plantations. In winter, typically on coastal grazing marshes.

Nightjar

The astonishing nighttime mechanical churr of the male Nightjar warrants an annual pilgrimage to heaths and forest clearings to hear it.

STAR SOUND: ADVERTISING CHURR At dusk, from May to late July, the males sing from elevated perches such as dead snags on trees. The sound is like the purring of a small engine; 2 minutes without a break is not unusual! It runs at 25 revs per second for 4–6 seconds and then switches to higher revs (but lower pitch) for a second, as if the little motor changes gear. The churr sometimes ends abruptly but often splutters to a muffled halt. It is the fifth sound on Track 6.

WING CLAPPING After a period of churring, males take flight, giving two to six sharp wing-claps at human hand-clap rate soon after take-off.

FLIGHT CALL Males in particular but also females give distinctive, rather pure and liquid *k'wip* or *p'wik* notes.

Caprimulgus europaeus

UK STATUS Summer visitor from central and southern Africa, with around 4,000–5,000 breeding pairs, mainly in southern England but very thinly spread elsewhere, even into southern Scotland.

HABITAT Most often associated with lowland heathland and clear-fell or newly planted conifer plantations, but also found on some moorland fringes.

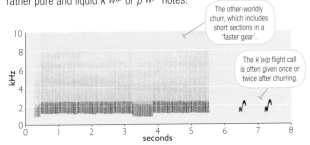

The other-worldly churr, which includes short sections in a 'faster gear'.

The *k'wip* flight call is often given once or twice after churring.

Swift

The aerial scream of this scimitar-winged master flyer is, for a brief few weeks, the essence of high summer near their breeding colonies. See page 82 for full description.

FLIGHT CALL Given in dashing group chases low over the rooftops, especially in the evening, it is a shrill, high-pitched *zreeee(t)*, either clear in timbre or with an audible buzz, many finishing with a slight drop in pitch at the clipped ending. Rarely calls when feeding or on migration.

WING-CLAP In flight, will sometimes clap their wings, either once over their backs or a few times first above and then below. Adults prospecting for nest sites will also bang at hole entrances, presumably to see if they are occupied.

Apus apus

UK STATUS Declining but widespread summer visitor, with around 80,000 pairs.

HABITAT Only lands to lay eggs and raise young, usually in holes in buildings under the eaves. Otherwise it is aerial, travelling long distances to feed.

Kingfisher

Few birds in Britain can match the visual appeal of the Kingfisher, with turquoise crown and wings, orange underparts and electric-blue rump, and yet it can be very hard to spot when perched in bankside vegetation. It means that learning the flight whistles will dramatically increase your chances of finding one.

FLIGHT CALL Given in direct bullet-flight low over the water and especially on take-off or before landing, the calls vary just slightly in pitch around a fairly high 6kHz, and include short, whistled *svee* notes and even shorter *tit* or *svit* notes. They are sharp enough to be heard above running water, with a hint of referee's whistle, and are often delivered in short, quick series. It is similar to the Dunnock's call but sharper, and often clearly from a fast-moving bird!

ALARM CALL In territorial disputes, gives fast *tididididit* calls, at the same high pitch as the main flight call.

ADVERTISING CALL Rarely heard, it strings its call notes into a half-hearted series, sometimes with a modicum of rhythm, such as *svit svee svit svee s'vit-s'vit s'vit s'vit....*

Alcedo atthis

UK STATUS Resident and short-distance dispersive migrant, with about 4,000–6,000 pairs across lowland areas, numbers crashing in harsh winters.

HABITAT Slow-moving rivers, lakes and gravel pits, where it digs deep nesting tunnels in vertical earth banks. Often perches motionless on choice branches overhanging water from which it can dive for fish. Outside the breeding season, many move downstream, often onto estuaries.

WHERE TO HEAR Viewable breeding banks include: Rainham Marshes (Essex/London); Rye Meads (Hertfordshire); Slimbridge (Gloucestershire); Lackford Lakes (Suffolk).

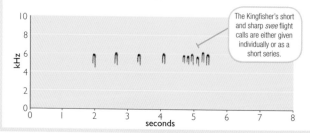

The Kingfisher's short and sharp *svee* flight calls are either given individually or as a short series.

Bee-eater

Starling-sized and in a painter's palette of blue, red, orange and green, the Bee-eater looks highly exotic. It is an expert flyer, often in upper airspace, and its frequent contact calls are a key way of finding them; indeed, many overflying birds are heard but are too high to see.

Merops apiaster

UK STATUS Rare annual visitor, mainly in spring, with up to 100 seen each year. There are signs that numbers are slowly increasing. Once every few years, a pair (or two) will stay to breed.

HABITAT Warm, dry places with plenty of bees; nests in burrows in sandy banks.

CALL A liquid, rolling *prrrip*, sometimes with shorter notes in between, such as *p' p' prrūip*....

Hoopoe

Its English and scientific names both derive from the male's laid-back advertising call. Mistle Thrush-sized, it has broad, rounded butterfly wings, boldly patterned in black and white, with a long, curved bill and erectable crest.

Upupa epops

UK STATUS Annual visitor, with around 100 records each year, mainly in spring.

HABITAT Likes feeding in short grass, and can sometimes turn up in gardens on lawns.

ADVERTISING CALL A delightful *oop-oop-oop* with a simple 1-2-3 rhythm, gentle and breathy but carrying quite far.

ALARM CALL On rising, sometimes gives a harsh, dry *raah*.

Wryneck

This relative of the woodpeckers has a character of its own, an ant-eater with a cryptically speckled and barred plumage. It became extinct as a regular breeder in southern Britain in the 1970s.

Jynx torquilla

UK STATUS Scarce migrant, mainly in early autumn, with up to 400 seen each year. Occasionally breeds in the Scottish Highlands.

HABITAT Coastal scrub with warm, dry, bare banks and short grass.

ADVERTISING CALL Now rarely heard in this country, it is a bold series of 15–25 *kew* notes, around five notes per second, typically rising in pitch and volume at the start. It is similar to the Lesser Spotted Woodpecker's call but lower pitched.

OTHER CALLS Occasional *teck* notes in alarm.

Great Spotted Woodpecker

By far the commonest woodpecker, Starling-sized and boldly pied; see page 96 for full description of the spring drumming and loud call.

DRUMMING A loud, short volley of rapid strikes on resonant tree trunks, typically for less than a second, totalling 10–20 strikes, perceptibly accelerating and become quieter towards the end.

CALL A simple, loud *kik!* or *tchik!*, very short and sharp.

ALARM CALL When agitated, gives a fast series of harsh *chet-chet-chet-chet...* or *chak-chak-chak-chak...* calls, loud, panicky, and often falling in pitch and slightly slowing at the end.

BEGGING NESTLINGS Older chicks in the nest can be very vocal, making an endless *si-si-si-si-si...* as they poke their heads out.

COMPARE WITH Lesser Spotted Woodpecker (drum and call).

Dendrocopos major

UK STATUS Sedentary resident throughout much of mainland Great Britain with perhaps 150,000 pairs, spreading north and a new arrival in eastern Ireland.

HABITAT Woodland, both deciduous and coniferous, where it drills nest holes in live or dead wood. Has adapted well to larger and leafier gardens.

Lesser Spotted Woodpecker

Only sparrow-sized, increasingly rare and often invisible high in the trees, knowing the sound of its drum, mainly given February–March, can be essential to locating the Lesser Spotted Woodpecker.

DRUMMING The rapid-fire drumming on branches, about 20 strikes per second, is longer than that of the Great Spotted Woodpecker, usually for 1–1.5 seconds, and each strike maintains the same intensity and pace instead of accelerating and fading away, so it is like a small mechanical drill. The sound is rather weak, and it sometimes gives two drums in quick succession.

ADVERTISING CALL A fast, high-pitched, rather sharp *week week week...* or *pee-pee-pee-pee...*, typically 8–15 notes at six to seven notes per second, higher and thinner than the similar but louder call of the Kestrel or the rare Wryneck; Nuthatch song is faster and more piping or reminiscent of a car alarm.

CALL A broad-spectrum, sharp *chik!*, much like the Great Spotted but higher pitched. In alarm, it is repeated in very fast, uneven series.

Dryobates minor

UK STATUS Sedentary resident, very thinly spread over lowland England and Wales, with only 1,000–2,000 pairs; only a few hotspots remain such as the New Forest (Hampshire).

HABITAT Largely ancient woodlands, especially with plenty of thin dead trees and branches, and preferring wet woodlands. Very rare in gardens – a woodpecker at a feeder will almost certainly be the Great Spotted.

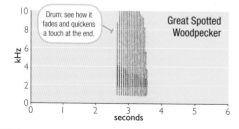

Drum: see how it fades and quickens a touch at the end.

Great Spotted Woodpecker

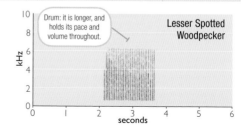

Drum: it is longer, and holds its pace and volume throughout.

Lesser Spotted Woodpecker

Green Woodpecker

The best laugh in the British bird world comes from this red-capped, moss-green, large woodpecker. See page 98 for full description.

'YAFFLE' ADVERTISING CALL A loud, attention-grabbing peal of 12–20 laughing notes, with a ringing, almost twanging timbre. The series often falls away a little in pitch and volume towards the end.

CALL Often given in flight or when flushed, it is typically two or three loud *kyak! kyak! kyak!* notes, sounding rudely startled. The call becomes full throttle if agitated or scared.

DRUMMING Males drum rarely, and when they do it is fast and weak.

COMPARE WITH Sparrowhawk; Kestrel.

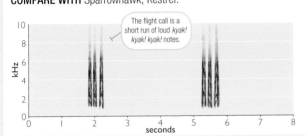

The flight call is a short run of loud *kyak! kyak! kyak!* notes.

Picus viridis

UK STATUS Resident and sedentary, mainly in the south and east, and absent from Ireland, with a breeding population of around 50,000 pairs. Rare in upland areas.

HABITAT Parkland, open woodland, old pasture, sandy heaths and a few large urban parks, with a mix of old trees and warm open ground with short vegetation, ideal for ants.

Kestrel

Best known for its accomplished hovering, the name 'Kestrel' derives from the French *crécelle*, meaning rattle, a reference to its call.

ALARM/EXCITEMENT CALL Often given when circling close to the nest site at the start of the breeding season, it is a fast *wik-wik-wik-wik…*, about seven notes per second. It may sometimes truncate these calls so that they are very clipped, separate *tik!   tik!…* notes. The calls of the Merlin and Hobby are confusingly similar.

SHIMMERING CALL Used in excitement when mates greet each other or pass food, this is a beautiful repeated *sirrrrrrrr*, each note about a second long and full of vibrato and rising in pitch.

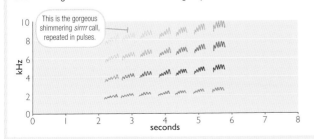

This is the gorgeous shimmering *sirrr* call, repeated in pulses.

Falco tinnunculus

UK STATUS Widespread but declining resident, largely sedentary, with a population of around 45,000 pairs; scarce in mountainous and urban areas.

HABITAT Hovers over rough, grassy habitats to catch voles and mice; in some areas, also hunts small birds, including the chicks of ground-nesting birds. Nests in barns, holes in trees or, increasingly, nestboxes.

Merlin

Compact and dashing, our smallest falcon is a bird of open landscapes. Its occasional calls are only likely to be heard on its remote moorland breeding grounds.

ALARM/EXCITEMENT CALL Occasionally gives a series of sharp notes, rather high-pitched, sometimes a slow *seeek seeek seeek* but typically around eight per second, and sometimes very fast indeed.

Falco columbarius

UK STATUS Scarce northern breeder, with around 1,000 pairs. More widespread in winter.

HABITAT Breeds on heather moors; otherwise on open marshes and downs.

Hobby

This rather dark summer falcon, with wings as rakish as a Swift, gives a typical series of falcon wickering calls.

ALARM/EXCITEMENT CALL Fairly vocal in and around the nest; the call is very similar to the Kestrel's, a series of *kew-kew-kew-kew...*, sometimes rather slow at three to four per second, sometimes twice as fast. Otherwise rarely heard.

Falco subbuteo

UK STATUS Summer visitor across much of lowland England and Wales, with now close to 3,000 pairs.

HABITAT Hunts over heaths, wetlands and farmland.

Peregrine

The dramatic upturn in fortunes of this powerful, anchor-shaped falcon means that its fierce calls are now heard over many city centres

ALARM/EXCITEMENT CALLS Can be vocal and loud on or near the nest, with two main calls. One is very long, peeved notes, with lots of rolled guttural 'r's and a real whininess, *karrrrrrk karrrrrk karrrrrk...*, repeated endlessly. The other is a faster series of harsh, scolding barks, *rark-rark-rark-rark*, no more than six per second so slower than in the smaller falcons. Both calls are as if with a sore throat and can include elements of yodel.

DISPLAY CALL A pair sometimes 'serenades' each other at the nest with a quiet *ker-CHIP*.

COMPARE WITH Kestrel; Hobby; Sparrowhawk.

Falco peregrinus

UK STATUS Largely resident with now more than 1,700 pairs spread widely around the coast and urban centres. There is some dispersal in winter, especially of young birds.

HABITAT Nests on ledges on vertical cliff faces on the coast and in quarries, but also often in special boxes erected in the human high-rise cliffscape.

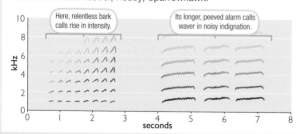

Here, relentless bark calls rise in intensity.

Its longer, peeved alarm calls waver in noisy indignation.

kHz — seconds

Ring-necked Parakeet

Urban myths claim that Britain's first Ring-necked Parakeets were released from the set of *The African Queen*, at Jimi Hendrix parties or by 1950s seamen returning from abroad. Whatever the truth, these Indian cagebirds are now flourishing in London and a few other sites, and proving well suited to city life. The call is often the first thing that alerts you to their presence, for they usually keep well hidden high in trees.

CALL Very vocal; the call is a penetrating, very loud, short shriek, *CHEEK!* or *SKREET!*, sometimes given singly but often in short or long series, at around four per second.

Psittacula krameri

UK STATUS Resident and sedentary, with more than 30,000 individuals and rising, so far mainly in London, Surrey and Kent.

HABITAT Needs tree holes for nesting and tall trees in which to gather at large communal winter roosts. Otherwise, mainly found in leafy suburbia feeding on fruit, nuts and at bird feeders.

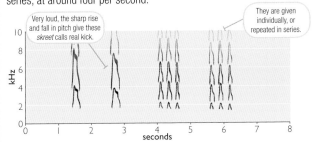

Very loud, the sharp rise and fall in pitch give these *skreet* calls real kick.

They are given individually, or repeated in series.

Red-backed Shrike

Once a widespread visitor in summer to dry heathlands in England and Wales, this masked bird, sometimes called the 'butcher bird' because of its habit of impaling its prey on thorns, was lost as a regular breeding bird in the 1980s but is still seen on passage.

CALLS In flight may give a *wrarr* or *charr*, in alarm a short, repeated *tchek*.

SONG Rarely heard; unmated males sing extended medleys of accomplished, if subdued, mimicry of other birds. To a female, however, he sings a fast twitter, not unlike a Linnet.

COMPARE WITH Lesser Whitethroat (call).

Lanius collurio

UK STATUS Scarce passage migrant, mainly in autumn, with up to 400 a year; very rare breeder.

HABITAT On passage, mainly in coastal scrub at migration hotspots.

Great Grey Shrike

This bandit-masked, monochrome bird watches for rodents and small birds in lonely vigil, perched up on bushes and tree stumps in its rather barren winter haunts.

CALLS Not at all vocal but can give simple calls, ranging from plain high whistles to ones with vibrato like a referee's whistle, but most typical is a drawn-out, Jay-like squawk, *whairk*.

Lanius excubitor

UK STATUS Scarce passage migrant and winter visitor from northern Europe, with about 200 seen each year.

HABITAT Mainly desolate heathlands.

Golden Oriole

Sadly, the small population in the poplar plantations of East Anglia petered out in the early 2000s, but a few continental birds still reach southern England each spring, giving you the slim chance of hearing the liquid song of the elusive, sunshine-coloured male.

SONG From a hidden perch high in a tree, each laid-back verse is a variant on a consistent theme, lasting about 0.75 seconds and sounding wonderfully tropical. It consists of about five deep, fluty notes, such as *dee-der-wee-dee-ō* (pause) *chee-oop-we'ee-ō* (pause).... It has an 'up and down' tune, with the final two notes often dropping.

FLIGHT CALL A retching, cat-like *re-er-eech*.

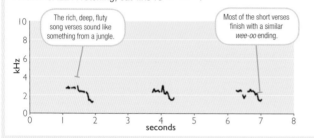

The rich, deep, fluty song verses sound like something from a jungle.

Most of the short verses finish with a similar *wee-oo* ending.

Oriolus oriolus

UK STATUS Rare and elusive spring migrant and occasional summer visitor, mainly to southern parts of England, as much inland as on the coast.

HABITAT Likes tall trees in waterside copses and woods.

Magpie

Track 48

Although it has long been a widespread bird, the reduction in persecution of the Magpie probably lies behind its increase in population, and it is now a familiar bird in most places, thanks to its bold, pied plumage, its cheeky behaviour, but also its loud calls. See page 87 for full description of its two main calls.

LONG RATTLE Its 5–20 loud *schak* notes rattle off at about 8–10 a second: *schak-ak-ak-ak-ak-ak-ak-ak*. It often repeats the rattle with only short pauses in between, and with different lengths of rattle each time. It is a percussive noise, with no tonal quality, given in alarm and in irritation.

SHORT CALL The conversational and much repeated *ker-chok!* and *ker-chik!* calls are contact calls. They have a little more audible pitch than the long rattle and so sound more musical. They are often repeated in prolonged chatter at its roost sites.

SONG Four song types have been described, which include Starling-like bleeps and whistles, interwoven with subdued *chok*-type calls, but they tend to be very quiet and poorly structured. They are only heard occasionally and are most often given in and around the nest.

COMPARE WITH Jackdaw (call); Mistle Thrush (call).

Pica pica

UK STATUS Sedentary resident across much of lowland Britain and increasingly in urban areas. It is still shot and trapped in large numbers in some areas but has an increasing population of around 600,000 pairs.

HABITAT Largely a bird of mixed farmland, where there are copses and thickets to breed and roost in. It is wary of people where still persecuted but increasingly tame in some urban situations.

Jay

This attractive woodland bird, salmon-pink with broad, rounded wings and white rump, is named after its distinctive, loud and harsh call.

CALL See page 99 for full description. Gives a loud, startling, harsh screech, *SCARCH!*, usually sparingly. Most are straight blasts of noise but a few have a hint of a musical note hidden among the rasp.

BUZZARD CALL It is an expert mimic of the mewing *myewwwwww* call of the Buzzard.

SONG The little-heard song is a half-hearted mutter, the notes rather random, quiet and disjointed with short clicks, squeaks, pips and bleeps and deeper *chop* and *charr* sounds. It can also include bird imitations, typically harsher calls such as those of the Great Spotted Woodpecker or the Magpie.

COMPARE WITH Buzzard (mewing call).

Garrulus glandarius

UK STATUS Widespread and mainly sedentary resident, with about 170,000 breeding pairs. In some winters, small numbers of continental birds arrive here.

HABITAT Generally a woodland bird, especially lowland oakwoods, but increasingly found in parks and gardens.

Chough

This red-billed, acrobatic crow is named after its exciting, reverberating call, which is often the first indication of its presence in its clifftop haunts. From its sound, it seems likely that the bird's name was originally pronounced to rhyme with 'bough' rather than 'rough'.

STAR SOUND: 'CHOUGH' CALL Vocal throughout much of the year, you sense that this is a bird that enjoys calling! The sound is fairly consistent, and is better written as *c^hyow!*, with a whizzing rise-and-fall, as when a child makes *p^eeow* sounds to imitate gunshots. The end of the call can sound heavily smudged: *c^hyowzh*. Overall, it is reminiscent of the Jackdaw's *chyar* but has more of a buzzing, electric feel. There are variants of the call but all have the same fizz. When given on the ground, birds bow slightly as they call, flicking their wings.

OTHER CALLS If really distressed, it can make harsher, scolding notes; occasionally a pair will give something akin to song, with low-volume warbles, squeaks or, more often, liquid *chiop* notes and barks.

These typical calls show the sharp drop in pitch – *chyow!*

Pyrrhocorax pyrrhocorax

UK STATUS Sedentary resident, with around 400–500 pairs, mainly in the far west of Wales, Scotland and Ireland. A small group of birds from southern Ireland naturally recolonised Cornwall in 2001 after more than 50 years' absence.

HABITAT Grazed coastal wildflower turf, damp enough for it to probe for beetles, especially under animal droppings. Requires cavities in cliff faces for nesting.

WHERE TO HEAR The Lizard (Cornwall); Welsh coastal clifftop paths; Isle of Man; Islay (Argyll); western Ireland.

Jackdaw

Often seen associating with Rooks or Starlings, or nuzzling up to its mate sitting atop a chimney, this small, silver-headed crow has calls that include some of the bird world's 'happy' sounds.

CALLS See page 86 for full description. Very vocal throughout the year; the three most frequently heard calls are staccato, upbeat *chyak!*, *chyar!* and *chyok!*, high-pitched for a member of the crow family. These are often repeated in excited little outbursts, inspiring others around them to join in. The combined noise from a pre-roost gathering, or from a colony racing in aerial circuits at dawn and dusk, can be very loud indeed.

ALARM CALL A longer, unhappy, straight, deep rasp, *charrrr*, 0.75 seconds long, more like a Rook call.

SONG Little heard, quiet and probably given most often by a pair in close contact, it is a medley of *chip* and *chap* sounds, sometimes with a rudimentary rhythm.

COMPARE WITH Chough; Carrion and Hooded Crows; Rook.

Corvus monedula

UK STATUS Widespread abundant resident, with around 1.4 million pairs, absent only in high mountains and the far north and west of Scotland.

HABITAT Mainly found in the farmed countryside, villages and especially parkland, feeding out on short turf. Tends to nest in small colonies on cliffs, quarries, ruins, churches and down open chimneys.

Raven

This mighty crow is almost Buzzard-sized, with a wingspan some 30cm larger than the Carrion Crow and with a hefty hatchet of a bill. However, at a distance, when it is difficult to gauge size or pick out the wedge-shaped tail, the characteristic calls can be a very useful identification aid.

CALL The typical flight call is a short, straight, hoarse and grating bark, very deep in pitch at around 1kHz and audible over large distances. It is sometimes called 'cronking', and is best rendered as *rrōk*, each call lasting only about 0.25 seconds. There are many variants, including ones lacking the bass croak and then sounding a little Crane-like, such as *RŌ-hu*. Sometimes a few notes are given in slow, uneven series.

ALARM CALL A faster, often uneven series of clipped notes, about four per second, such as *brōk! brōk!....*

SONG As with most of the crow family, the song is rudimentary and almost apologetic, only likely to be heard at close range and sounding like an embarrassed practice session.

COMPARE WITH Carrion Crow; Hooded Crow.

Corvus corax

UK STATUS Sedentary resident across most of western and northern Britain, and making steady inroads into its former range across central lowland areas, but still scarce in the east. The population is around 7,500 pairs and rising.

HABITAT Until recently, was found almost wholly in wild cliff and mountain scenery, but it is becoming more familiar again in lowland farmland. Some nest on cliffs, others build large tree nests. The breeding season starts early, with eggs often laid in February.

Rook

The Rook and Carrion Crow can be rather difficult to tell apart visually at any distance, and especially in flight overhead, which is where the subtle difference between their calls can help. However, the sound of a rookery is enjoyable just in itself, one of the defining sounds of the countryside.

CALL See page 112 for full description. Although the basic cawing call is variable, it is typically an *arrrrr* or *corr*, level or gently falling in pitch, in contrast to the Carrion and Hooded Crow's deeply downslurred *raarrrrk*. Importantly, Rooks don't tend to repeat it in regular series, unlike the crows, and the heavy rasp running through the Rook's call is more pleasant and 'sung' than the aggressive, rough feel of the crow.

ROOKERY CALLS It can be difficult to make out individual calls among the cacophony of a rookery. However, amid the basic chorus of loud *corr* notes and their variants as adult birds bow deeply to each other, it is possible to hear twice-as-high *brrup* notes, gurgles and rasps, which is a primitive song. As the season progresses, the squeaks and then caws of begging youngsters add to the mix.

COMPARE WITH Carrion Crow; Hooded Crow.

Track 79

Corvus frugilegus

UK STATUS Widespread and abundant resident as far north as Shetland in all but urban centres, mountains and smaller islands, with about 1.1 million breeding pairs.

HABITAT Mostly farmland, where it feeds on the ground on pasture and ploughed land. Requires copses and woods for its dense communal colonies ('rookeries'). A few have taken to visiting garden feeders, where some have learnt to cling to feeders, if in an ungainly way.

Carrion Crow and Hooded Crow

It was not until 2002 that these were finally recognised as two different species, despite them looking so distinct, the Carrion (far right) all black and the 'Hoodie' (right) with its neat grey 'waistcoat'. However, there is little to tell between their calls.

CALL See page 85 for full description. The main call is a harsh, loud, deep *raarrrrk* or *karrr*, noticeably dropping in pitch. It sounds malevolent, and unlike the Rook's is typically repeated mechanically three or four times. Some variations sound gargled or strangled.

RAPTOR ALARM This useful call usually signals that a crow is mobbing a flying raptor, and is typically a higher-pitched *arrrp arrrp*, with excessive rolling of the 'r's.

SONG Occasionally gives a rudimentary, introspective, quiet song in spring, with a range of blips and caws, including bill-clicking.

COMPARE WITH Rook; Raven.

Track 46

Corvus corone
Corvus cornix

UK STATUS Sedentary residents, the Carrion Crow with about a million pairs, the Hooded about 260,000. 'Hoodies' are found in Ireland, the Isle of Man, and north and west of a line from Arran to Inverness, the Carrion everywhere else.

HABITAT Cosmopolitan, from farmland to urban centres to beaches.

Waxwing

Everyone wants to see a Waxwing, Starling-sized yet salmon-buff in colour with a long, wispy crest, heavy black eyeliner, and bright yellow tips to the tail and flight feathers. Befitting this midwinter 'gift' from Lapland, the calls have an intriguing flavour of Christmas!

SLEIGH BELLS CALL Like surging pulses of faint jingling sleigh bells, the trills are a very high-pitched '*sirilee sirilee...*', easily overlooked in the urban soundscape. There is a hint of short bursts of distant Grasshopper Warbler song. In flocks, the combined calls seem to surge and fade, and you can hear the slight differences in pitch between individual birds.

BEAK SNAPPING Resting flocks can make quiet but audible 'keep your distance' bill snaps at each other.

SONG Occasionally heard in the UK on sunny, still days from the few birds that linger into late April or May, it is barely more than slightly extended call trills, sometimes sliding in pitch, sometimes interspersed with harsher notes.

Bombycilla garrulus

UK STATUS Annual winter visitor in varying numbers, with a few hundred each year, but in occasional irruptions many thousands visit.

HABITAT Seeks out any tree or shrub still holding berries and fruit, so comes into towns and cities to find pyracanthas, cotoneasters, crab apples and rowans; they are often found in supermarket car parks!

WHERE TO HEAR Favoured cities include Nottingham, Newcastle and Aberdeen, but they are possible in any urban area in a good year.

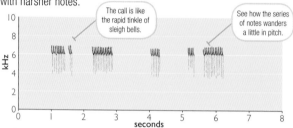

The call is like the rapid tinkle of sleigh bells.

See how the series of notes wanders a little in pitch.

Coal Tit

Tracks 25 & 29

This dainty tit is a key voice of conifer trees, sounding like the cute baby cousin of the Great Tit. See pages 68 and 71 for full description.

SONG Follows the same basic song pattern of the Great Tit, with a series of seesawing couplets or triplets, the key differences being the higher pitch and the thinner, weedier timbre, often sounding like the notes are warped. Sometimes the pacing is faster than that of the Great Tit.

CALL Has a wide range of conversational calls, which all tend to have the same high pitch and distinctive timbre as the song, being rather cute, baby-like and 'warped'. Typical of the bendy notes are $s_i\bar{u}$, $ts_{\bar{u}}$ or $t^{s\bar{u}}$, plus a deeper *chip d'dip*, while some *si-si-si* calls are so thin and high that they are easily confused with those of the Goldcrest.

COMPARE WITH Great Tit (song and call); Blue Tit (call); Goldcrest (call).

Periparus ater

UK STATUS Sedentary resident, with around 750,000 pairs. A few continental birds arrive in some autumns.

HABITAT In the breeding season, found in conifer trees, from forestry plantations to churchyard yews and garden cypresses. Nests mainly at ground level and does not take readily to nestboxes. In winter, seen in ones and twos at garden feeders.

Crested Tit

A pilgrimage to the magnificent pine forests of Highland Scotland is the only way to see this special bird in the UK. Even then, it can still take some finding, so knowing the distinctive fast shuttling chatter is essential.

CALL A simple series of four to seven notes, delivered as a fast shuttling chatter, *twidididit*, at a rate of about 15 notes per second. Each run of notes wanders a little in pitch, often with a slight rise-and-fall. It is fairly vocal early in the breeding season, but be aware that it can be silent for much of the summer.

SONG Unsophisticated, all it tends to do is give the main call a couple of times and then alternate with a short series of higher-pitched s^{wee} s^{wee} s^{wee} s_i notes. Learning the basic call is the key to recognising the song.

COMPARE WITH Long-tailed Tit (*sirrut* call); contact calls of other tits.

Lophophanes cristatus

UK STATUS Sedentary resident, with birds rarely straying more than a few miles from where they hatched, and with a population of 1,000–2,000 pairs.

HABITAT Feeds among the needles of Scots Pine trees and in large sunny patches of mature heather between the trees; in winter on bird feeders nearby.

WHERE TO HEAR In the Highlands, Loch Garten and Forest Lodge (Abernethy Forest), Loch an Eilein (Rothiemurchus Forest), Corrimony (Glenurquhart) and Glen Affric. Also Culbin Forest (Moray).

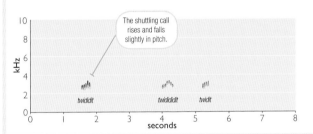

The shuttling call rises and falls slightly in pitch.

twididit

twidididit

twidit

Great Tit

Tracks 23 & 28

Its signature song is a strident 'teacher teacher' but with many variations, which with its calls add up to a very wide repertoire. See pages 66 and 70 for full description.

SONG At its simplest, it gives a seesawing couplet five or so times in a row to a metronomic beat, but the many variations include jauntier rhythms on the two-note theme, or triplets. All have a characteristic bold, ringing timbre.

ALARM CALL A simple, dry rattle, about 12 notes per second, steady and uniform, like a fast mini-Magpie or shake of the maracas: *chrrrrrrrrt*.

PINK PINK CALL A cheerful, sharp *pink pink*.

BEGGING CALL Fledged young give little runs of tinny *tzi tzi tzi* notes, usually rising in pitch.

CONTACT CALLS Varied, ranging from mimicry of other birds to soft tweeting contact calls, plus clipped *pee*, *tit*, *chit* and *tsee* notes.

COMPARE WITH Coal Tit (song and call); Blue Tit (call); Chaffinch (call).

Parus major

UK STATUS Sedentary, abundant resident, with about 2.5 million pairs, found across the British Isles except the Outer Hebrides, Orkney and Shetland.

HABITAT Needs trees, so found in all types of woodland but also parks and large gardens, and is a regular at bird feeders.

Marsh Tit

Track 66

Another of our declining woodland species, the calls are crucial for separating the Marsh Tit from the Willow Tit.

SNEEZE CALL A sharply dropping *pit-choo* or longer *pitchi-choo* or *tsi tsi choo*. It is usually given singly. See page 101 for full description.

BEE-BEE-BEE **CALL** A short series of buzzing, nasal, angry-sounding deep notes, *bee-bee-bee*, often with one or two high introductory notes, *chikka-bee-bee-bee*.

SONG Infrequently heard; each male has several different verses, usually a fast repetition of one note about 10 times, such as *tchup tchup tchup...*, each note dropping so sharply it sounds whipped. Some variants are slower and some are of two syllables, such as *s'wee s'wee s'wee...*, *t'sū t'sū t'sū...* or *si-chū si-chū si-chū....*

OTHER CALLS As with other tits, it gives high contact calls.

Poecile palustris

UK STATUS Sedentary resident throughout much of lowland England and Wales, with about 40,000 pairs; still relatively common in some southern areas but decreasing in most parts.

HABITAT Largely a bird of mature woodland, requiring pre-existing holes to nest in; a few are seen in more rural gardens in winter.

Willow Tit

Poecile montana

Almost identical to the Marsh Tit visually, the slow whistled song and especially the deep scolding calls can be vital identification clinchers.

TCHAY CALL This key call is a harsh, nasal *tchay* or *tcharr*, each note drawn out for about half a second and sounding incredibly peeved. It is given singly or sometimes three or four in slow series, and can be introduced with high-pitched *tsi-tsi* notes.

SONG A simple but little heard *pyū pyū pyū...* or *tyū tyū tyū...*, the pure whistled notes with a strong downslur, repeated four to six times, at a slow rate of three per second. Compared with the *pyū* song of the Wood Warbler, it has fewer notes and each note drops more sharply in pitch. Marsh Tit song is occasionally based on a similar note but is much faster.

OTHER CALLS Has a small range of typical tit contact notes, high-pitched, often just a simple *si-si*.

COMPARE WITH Marsh Tit (song and call); Wood Warbler (song).

UK STATUS Very sedentary resident with perhaps only 3,000 pairs remaining, mainly in Wales and central and north-east England, with a few in southern Scotland.

HABITAT Wet woodland where it excavates nest holes in rotting tree trunks. It is rare at garden feeders.

WHERE TO HEAR Try: Cors Caron (Ceredigion); Carsington Water (Derbyshire); Risley Moss (Cheshire); Pennington Flash (Greater Manchester); Old Moor (South Yorkshire); and Fairburn Ings (West Yorkshire).

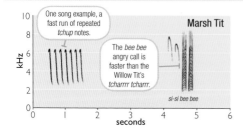

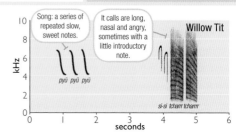

Blue Tit

Although very familiar at winter bird tables, its song and alarm call are much less known, but will help you find it when the leaves are on the trees. See pages 66 and 70 for full description.

SONG Very varied, but each verse usually lasts just 1.5 seconds, and the key is structure. Most start with two or three very high introductory notes and step down into a lower, straight-pitched, rapid trill, typically *sispi si-hi-hi-hi-hi*. Some variants have a slower trill, others are simpler, such as *tsi-tsi-sup-sup*, but the 'intro, drop low' pattern is almost universal.

ALARM CALL Very like the 'mini-Magpie' rattle of the Great Tit, but often faster and tending to rise in pitch, *chrrrrrᵗ*.

OTHER CALLS A wide range of little calls, as with other tits, some as high and thin as Goldcrest calls, some stronger, but the 'high-drop-to-low' theme often remains, such as *tsi-tsi-sup* or *tsi-zee-zee*.

COMPARE WITH Great Tit (alarm call, call); Treecreeper (song).

Cyanistes caeruleus

UK STATUS Abundant and sedentary resident, with around 3.5 million pairs spread across almost all of the British Isles.

HABITAT Needs mature trees throughout the year, preferably deciduous. It is most abundant in ancient oak woodland but also frequent in leafy suburbs, timing its nesting to coincide with hatching moth caterpillars.

Bearded Tit

The distinctive calls are often your best way of locating this long-tailed and rather tit-like bird which tends to stay well hidden in reedbeds.

CALLS Frequently heard from deep in the reeds, imagine how a doll's cash register might sound, or a tiny typewriter at the end of a line: *ching!* Some calls lack the twang, being more of an unremarkable *tip* or even weaker *chu*, but pretty soon the pinging note should be heard. Each call is uttered singly, not as a quick series, but troops calling to each other produce a conversation.

SONG There is a rudimentary song but it is rarely heard, a series of simple harsh and twanging notes, and in particular a buzzing *brrrrr!* like that of a Lesser Redpoll.

COMPARE WITH Lesser Redpoll (song).

Panurus biarmicus

UK STATUS Largely sedentary resident, with a scattered population of only 600–800 pairs, mainly around the East Anglian coast and fens and in the Tay Reedbeds (Perthshire), but some along the south English coast, Yorkshire and elsewhere, with eruptive movements at the end of summer.

HABITAT Reedbeds, usually large ones.

WHERE TO HEAR Try: Ham Wall (Somerset); Minsmere, Walberswick and Lakenheath Fen (Suffolk); Cley and Titchwell Marsh (Norfolk); Leighton Moss (Lancashire); Blacktoft Sands (East Yorkshire).

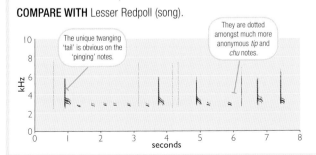

The unique twanging 'tail' is obvious on the 'pinging' notes.

They are dotted amongst much more anonymous *tip* and *chu* notes.

Woodlark

This rather anonymous, streaky brown bird found on bare heathy ground and open forests has an exquisite song and jaunty flight call.

STAR SOUND: SONG From January, it is delivered from an open perch or in high-circling display flight. Each verse lasts about 5 seconds and is a series of sweet couplets dropping gently down the scale, seesawing all the way down and accelerating. Each verse uses a different couplet and rhythm, like someone finding various ways to skip lightly down the stairs. It is gloriously inventive and a joy to listen to.

CALL Typically three bouncing syllables or half-syllables in many different versions, such as $t^u{}'l'w_it$, $j^e\ t'lui$, $w^e{}_{e-er}'w_it$. It has all the lilt and charm as if it were said with a French accent.

Lullula arborea

UK STATUS Scarce resident in southern England, East Anglia, the Midlands and north-east England with about 3,000 breeding pairs. They disperse locally in winter.

HABITAT Breeds on lowland heathland with scattered trees, and also in clear-felled conifer plantations, but in winter may move onto arable fields.

WHERE TO HEAR Heaths of Hampshire, Sussex and Surrey; Thetford Forest and neighbouring heaths (Norfolk); Suffolk Sandlings.

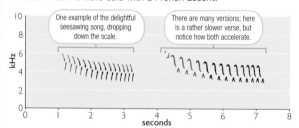

One example of the delightful seesawing song, dropping down the scale.

There are many versions; here is a rather slower verse, but notice how both accelerate.

Skylark

Track 80

High above open fields, this amazing songster hangs as a dot in the sky and lets its endless song roll. See page 113 for full description.

STAR SOUND: SONG One verse can be unbroken for many minutes, a burbling doodle with frequent quick repetitions of individual notes, short trills and short phrases. Around half the notes have a pleasant, rolling burr, and it can interweave short bursts of mimicry.

FLIGHT CALL When flushed or on migration, it has a range of rippling, cheerful calls such as a double- or triple-syllable *syrup* or *chir'r'rup*.

Alauda arvensis

UK STATUS Widespread, abundant but much declined resident, with around 1.5 million pairs. Also a winter visitor and passage migrant.

HABITAT Open habitats, especially extensive farmland but also dunes and moors; in winter feeds in weedy stubbles and saltmarsh.

Shore Lark

This can be an inconspicuous bird, creeping around silently in a winter saltmarsh. On taking flight, its rather soft flight call becomes useful.

FLIGHT CALL Typically gives high-pitched, short and rather feeble whistles, *sih*, or with a bit of ripple, *si-li-li* or *s'lit*, but lacking the confident, well-expressed *syrup* of Skylark.

Eremophila alpestris

UK STATUS Scarce winter visitor, with about 100 seen each year.

HABITAT Mainly vast open saltmarsh.

Sand Martin

Seen well, the sandy-brown plumage and lack of white rump separate it from the House Martin, as does its habit of nesting in colonies in river banks and sandy cliffs, but the 'dry as sand' call is a useful extra clue.

CALLS Very vocal around the nesting colony but quieter elsewhere; the main call is a shuttling, dry *cht-cht-cht*…. It is a noise as desiccated as the sandy banks it nests in, without any obvious pitch. When excited, the notes run into a fast, pulsing buzz, at a rate of about 10 notes per second: *j-j-j-j-j-j-j-j-jt*.

RAPTOR ALARM CALL On seeing a bird of prey such as a Hobby, it makes loud *eerrp!* or *cheerrtt!* notes, more musical than its main call.

COMPARE WITH House Martin (call).

Riparia riparia

UK STATUS Summer visitor and passage migrant from sub-Saharan Africa, some arriving by March. Colonies are often transient and the population is difficult to estimate, with maybe 50,000–200,000 pairs.

HABITAT Needs vertical sand or soft earth banks in which to excavate nest tunnels, such as along rivers, coastal sand cliffs, and sand and gravel pits. Typically feeds over the water and roosts in reedbeds on migration.

Swallow

Track 86

The sweet, fast, twittering song is a pleasant and upbeat sound of summer. See page 119 for full description.

SONG A fast twittering at a relentless rate of about eight 'mashed-up' notes per second. Towards the end of a verse, it throws in an incongruous, drawn-out 'snairrr' note, followed by a dry ratchet, *prrrrrt*.

FLIGHT CALL A brief, cheery *wit!* or *wit-wit!*

RAPTOR ALARM CALL A sharp, whipped *ser-wit!*

COMPARE WITH House Martin (song); Goldfinch (call).

Hirundo rustica

UK STATUS Very common summer visitor, April–October, with around 800,000 pairs, rare only in the far north and west.

HABITAT Farmland, nesting in barns and outhouses and hunting for insects in flight low around livestock and over lakes and reservoirs.

House Martin

Track 41

Their distinctive call helps separate fast-flying House Martins from Sand Martins and Swallows, while the song is best heard at their mud-cup nests. See page 81 for full description.

FLIGHT CALL A rather lisping *preet*, or two-syllable *pritit* or *chirrit*, not as 'dry' as the Sand Martin's call.

ALARM CALL A loud *sirrrt!*

SONG An enthusiastic, long, mushy babbling, with the *preet* note woven in.

COMPARE WITH Sand Martin (call); Swallow (song).

Delichon urbicum

UK STATUS Widespread but declining summer visitor, mainly mid-April–early October, with around 500,000 pairs nesting in small, loose colonies.

HABITAT Nests under the eaves of buildings where there is a clear flight line in. Otherwise, catches insects high in the sky, or low over lakes and reservoirs when the weather is poor.

Cetti's Warbler

Track 96

Notorious for its elusiveness, this bird makes its presence known with the most explosive of songs.

STAR SOUND: SONG See page 128 for full description. A male sings his verse just once or twice from a song

post before moving to the next in his territory. It is best remembered as 'CHIP!   CHIP-PEE!   CHIPPY-CHIP-SHOP CHIPPY-CHIP-SHOP'.

CALL An abrupt, hard *chit!* that can run into uneven bursts, or even into an extended fast, rattling chatter.

Cettia cetti

UK STATUS Sedentary and increasing resident in southern lowland England and Wales. The population can crash in hard winters but is around 2,000 breeding pairs.

HABITAT Wetland margins, where dense vegetation such as willows and brambles line the edges of pools, streams and canals. Often moves into reedbeds in winter.

Long-tailed Tit

Track 68

The sound of a Long-tailed Tit troop is a constant delightful chatter. See page 103 for full description.

***SIRRUT* CALL** When a flock is anxious, it gives repeated *sirrut* calls, strongly downslurred with rolling 'r's.

***SI-SI-SI* CALL** High-pitched, it is typically given in threes. In alarm, it can be faster, longer and louder: *si-si-si-si-si-si-....*

***TUC* CALL** A quietly conversational call, just *tuc* or *tup*.

SONG Rarely heard, and not used for marking a territory, it is a speeded-up, subdued montage of its various calls.

COMPARE WITH Tit species (calls); Goldcrest (call); Treecreeper (call).

Aegithalos caudatus

UK STATUS Sedentary, widespread resident across almost all of the British Isles, with an estimated 340,000 territories. Populations crash during cold winters.

HABITAT Mainly a woodland bird, using hedges, reedbeds and other tall vegetation to move from one patch of trees to the next, and nervous of crossing open spaces.

Willow Warbler

Tracks 55 & 57

So similar in appearance to the Chiffchaff, the Willow Warbler's sweet, lilting song is an identification clincher. See pages 91 and 93 for full description.

STAR SOUND: SONG This beautiful lilting cascade is about 15–20 whistled notes, dropping unevenly down the scale.

CALL The soft *hoo-weet*, with an upwards inflection, is very similar to the Chiffchaff's *hweet*, but is two syllables and more wistful.

COMPARE WITH Chaffinch (song and call); Chiffchaff (call); Redstart (call).

Phylloscopus trochilus

UK STATUS Abundant and widespread summer visitor and passage migrant, with about 2.4 million pairs, the highest densities in the north and west.

HABITAT A rural bird of open, sunny woodlands and young plantations.

Chiffchaff

Spring starts when the first Chiffchaff arrives and sings its name enthusiastically. See pages 90 and 93 for full description.

SONG A simple series of clipped *chiff chaff*-type notes to a metronomic rhythm. The song is sometimes introduced with a soft hiccup, *h'ric*.

HWEET CALL A fairly soft, single-syllable whistle, *hweet*.

SEE-UP CALL In early autumn, juveniles give a *see-up* or *swee-oo* call.

CALL OF *TRISTIS* SUBSPECIES Small numbers of the pale Siberian subspecies winter in the UK – their call is a whistled *tsoo*.

COMPARE WITH Willow Warbler (call); Chaffinch (call); Redstart (call); Nightingale (call).

Phylloscopus collybita

UK STATUS Abundant and increasing summer visitor, with about 1.2 million pairs. Also a scarce but increasing winter visitor, with a few thousand individuals especially in the south and west.

HABITAT Breeds in deciduous woodland with ground cover for nesting. On migration, visits willows and reeds. In winter, especially found at sewage farms and wetland margins.

Wood Warbler

The male's two very different songs are a delightful feature of northern and western oakwoods. See page 108 for full description.

STAR SOUND: SPINNING COIN SONG The 3–5-second verses are a series of *sip* notes, starting slowly and accelerating like a spinning coin.

PYŪ PYŪ PYŪ SONG A simple, gentle series of eight or so melodic *pyū* notes, dropping slightly in pitch from start to end.

CALL A simple *pyū*, dipping slightly in pitch.

COMPARE WITH Willow Tit (*pyū* song).

Phylloscopus sibilatrix

UK STATUS Scarce and declining summer visitor, with around 6,500 pairs, widespread only in Wales and western Scotland.

HABITAT A specialist of ancient oakwoods in mainly upland valleys.

Yellow-browed Warbler

This tiny 'waif' from northern Russia, once a major rarity, is turning up in increasing numbers each autumn.

CALL The simple and consistent call is *se-ū-weest*, easily confused with the Coal Tit's call. The strong dip in pitch is noticeable.

COMPARE WITH Coal Tit (call).

Phylloscopus inornatus

UK STATUS Passage migrant in autumn, mainly mid-September to the end of October, with more than 1,000 in total in a good autumn. About a dozen winter each year, mainly in south-west England.

HABITAT On migration, especially fond of Sycamore trees or stands of willow.

WHERE TO HEAR Migration hotspots such as Scilly, west Cornwall, north Norfolk, Flamborough Head (East Yorkshire), and Shetland.

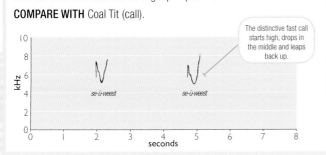

The distinctive fast call starts high, drops in the middle and leaps back up.

Sedge Warbler

The key to telling the songs of this wetland dweller from the Reed Warbler is the rhythm.

STAR SOUND: SONG See page 127 for full description. Each verse is a mix of harsh churring and sweeter notes that can continue unbroken for a minute or more. Unlike the 'Steady Eddie' Reed Warbler, the Sedge is a jazz singer! A verse switches from one rhythm to another, taking one phrase and repeating it, before changing tack and pace.

CALLS Variable and intergrading, they include a sticky, deep *thik* or *chuk*, rather like the Garden Warbler when given individually, but also run into a rattle of four to seven notes, often slowing at the end. It also gives fast pulses of clockwork-toy notes, rather like a Wren call.

COMPARE WITH Reed Warbler (song and call); Garden Warbler (call); Wren (call).

Acrocephalus schoenobaenus

UK STATUS Widespread and common summer visitor from Africa, with just under 300,000 territories. Commonest in coastal areas and lowlands.

HABITAT Breeds in damp, open habitats with dense vegetation, often with some bushes, such as in lowland fens and marshes.

Reed Warbler

This plain brown warbler is the 'Steady Eddie' of the reedbed compared with the jazzy Sedge Warbler.

STAR SOUND: MAIN SONG See page 126 for full description. Each verse is a series of harsh grating churrs and sweeter notes, up to 3 minutes long. Most of the time, you can beat your finger to every note in a steady '1, 2, 3, 4…' rhythm. It's not perfect, especially as verses start, when it can do excellent mimicry, but it will almost always settle into the beat.

CONVERSATIONAL SONG Infrequently, a male gives a subdued warbling version of his main song with less repetition.

CALLS Most typically and distinctively gives a dry *chrrrrk*, much longer and with faster rolling of the 'r's than the Sedge Warbler's call, or a shorter version of that, *chrek*.

COMPARE WITH Sedge Warbler (song and call); Whitethroat (alarm call).

Acrocephalus scirpaceus

UK STATUS Common summer visitor (April–September) and passage migrant with around 130,000 pairs, mainly in south, east and central English lowlands and south Wales, and expanding its range west and north.

HABITAT Almost wholly tied to reeds, whether in large reedbeds or linear stands such as along ditches and canals.

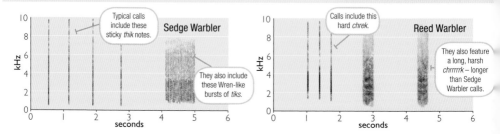

Sedge Warbler. Typical calls include these sticky *thik* notes. They also include these Wren-like bursts of *tiks*.

Reed Warbler. Calls include this hard *chrek*. They also feature a long, harsh *chrrrrk* – longer than Sedge Warbler calls.

Marsh Warbler

Visually very similar to the Reed Warbler, this is such a rare breeding bird that, sadly, there is little chance of hearing its incredible song.

SONG The most astonishing mimic, stringing together bursts of the songs and calls of a wide variety of European and African birds. If you are lucky enough to hear one, you can't help but marvel at its skill.

Acrocephalus palustris

UK STATUS Very rare breeder in the south-east and Shetland, and rare passage migrant.

HABITAT Rank vegetation in damp areas.

Grasshopper Warbler

Track 97

Acting more like a mouse than a bird and staying deep in thick cover, views of this bird are few and far between, but it compensates with one of the nation's strangest bird songs.

STAR SOUND: SONG See page 129 for full description. The reeling song is indeed like a grasshopper, at a constant, fast 25 notes per second. It is rather high-pitched, and verses are long, usually about 20 seconds but exceptionally an hour, with barely a pause.

CALL A short, clipped, occasional *tip! tip!*, rather like a Robin call, sometimes a little 'thicker', such as *teck*.

COMPARE WITH Savi's Warbler (song); Robin (call).

Locustella naevia

UK STATUS Fairly common but declining summer visitor and passage migrant, with around 16,000 pairs. It is commonest in west Wales, west Scotland and Ireland.

HABITAT Breeds in damp areas with rather dense, rank vegetation.

Savi's Warbler

If it was not for its distinctive song, very few of these rare skulking warblers would be seen. However, when one sets up territory, it may sing over several weeks.

SONG Like Grasshopper Warbler song but the notes are twice as fast, 50 per second, creating a buzz rather than a reeling noise, in which it is hard to make out the individual notes. Also the main pitch is deeper.

CALL A sharp *zwit*.

COMPARE WITH Grasshopper Warbler (song).

Locustella luscinioides

UK STATUS Rare summer visitor, breeding most years with usually only around five singing males in the whole of the country.

HABITAT Extensive reedbeds.

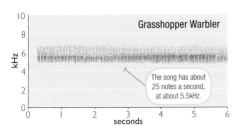

Grasshopper Warbler

The song has about 25 notes a second, at about 5.5kHz.

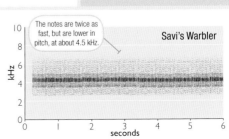

Savi's Warbler

The notes are twice as fast, but are lower in pitch, at about 4.5 kHz.

Blackcap

Tracks 56 & 57

In spring, our woodlands fill with the Blackcap's tuneful 'song of two halves'. See pages 92 and 93 for full description.

MAIN SONG Each warbling verse is about 3–5 seconds long, and starts hesitantly but breaks into a louder, pure, fluty melody.

LONG SONG This rambling warble, up to 30 seconds long, is full of squeaks and scratches and is very similar to Garden Warbler song, but without the latter's rippling or rich deep notes.

CALL A clean, hard *tak*, often in series.

COMPARE WITH Garden Warbler (song and call); Robin (call); Lesser Whitethroat (call).

Sylvia atricapilla

UK STATUS Abundant, widespread and increasing summer visitor, with more than 1.2 million territories. It is also an increasing winter visitor, mainly in the south and west.

HABITAT Breeds where there are mature trees, deciduous and mixed, even in urban parks, but it needs undisturbed ground cover for nesting. Wintering birds favour gardens.

Garden Warbler

Track 58

Learning to tell the song of the Garden Warbler from that of the Blackcap is difficult, but perseverance pays off … usually!

MAIN SONG See page 94 for full description of this energetic, 3–8-second warble, flowing like a babbling brook. It is the 'same song' throughout, in contrast to the Blackcap's typical 'unsure start, confident ending'. It often returns time and again to richer, deep, Blackbird-like notes, and there are generally only short pauses between verses.

CALL Similar to Blackcap's *tek* but a deeper, stickier *tchek*, usually delivered in stuttering series.

COMPARE WITH Blackcap (song and call); Lesser Whitethroat (call).

Sylvia borin

UK STATUS Common summer visitor and passage migrant, with around 170,000 pairs; widespread but very rare in Ireland, northern Scotland and urban areas.

HABITAT Poorly named, being rare in gardens and much more a bird of scrubby woodlands and thickets with dense ground cover, tending to avoid high forest.

Lesser Whitethroat

Track 85

The rattling song of this sooty-masked warbler is a vital way of finding it in its high hedgerow haunts.

SONG See page 118 for full description. It is a straightforward rattle of about 10–12 identical notes, a very fast *chy'ok chy'ok chy'ok*, rather mechanical in feel, lasting about 1.2 seconds, and often introduced with a few quiet, squeaky notes.

CALL A short *tjik*, similar to the Blackcap but a little 'stickier', and not as deep or long as the Garden Warbler's call.

COMPARE WITH Cirl Bunting (song); Blackcap (call); Garden Warbler (call).

Sylvia curruca

UK STATUS Fairly common summer visitor to mainly lowland England, with just under 75,000 pairs, expanding its range into coastal Wales and Scotland. Passage migrants turn up along our coasts.

HABITAT Tall, mature hedgerows and scrub thickets.

Whitethroat

This common but rather anonymous-looking warbler is heard in spring singing from hedges and bramble patches across the country, sometimes flying up to give a longer, sweeter song flight.

SHORT AND LONG SONGS See page 117 for full descriptions. Mostly gives short, 1–1.5-second, scratchy verses from within a hedge or bush, changing each time but with a consistent theme that lurches in pitch, high to low and back again two or three times. Sometimes, the male flutters up above his territory, singing a longer 3–5-second verse, like the 'short song' but with greater flow and sweetness.

ZREE **CALL** Clearly upslurred, the buzzy *zree* is often given in stuttering series.

ALARM CALL Often heard, a straight, irritated *chrrrr!*, rather long and insistent, repeated every second or so.

COMPARE WITH Dartford Warbler (song); Garden Warbler (song); Stonechat (song); Dunnock (song); Reed Warbler (call).

Sylvia communis

UK STATUS Abundant and widespread summer visitor and passage migrant from sub-Saharan Africa, with more than 1 million pairs, rare only in the far north. Although slowly recovering, the population is still less than half what it was in the 1960s.

HABITAT A specialist of dense low scrub and hedgerows in the lowlands; rare in urban areas.

Dartford Warbler

This hyperactive specialist of southern heathland often only gives brief views, so its super-speed scratchy song and jarring calls can be an important indicator of its presence.

MAIN SONG The fastest, scratchiest, most jaggedy warble, with energetic 1–2-second outbursts, like a breakneck Whitethroat. It mixes rather high-pitched notes with a core of deep, fast, rasping churrs. It is given from favourite song perches, often on gorse bushes, the male looking around as he sings before flitting to the next perch.

SONG-FLIGHT SONG This much longer version of the male's song is rather more flowing and slightly less harsh, with a hint of Dunnock song, as he flutters jerkily up several metres above the vegetation.

CALL A discordant, petulant *tcharr!*, sometimes followed by *tut*.

Sylvia undata

UK STATUS A scarce resident, with more than 3,000 pairs, mainly in southern England but now also in East Anglia, south-west England, coastal Wales and the Midlands. Some birds disperse in autumn to colonise new sites, and numbers crash in harsh winters.

HABITAT Almost exclusively a bird of gorse and heather on lowland and coastal heathland.

WHERE TO HEAR Aylesbeare Common (Devon); Arne (Dorset); the New Forest (Hampshire); the Wealden Heaths and Thames Basin Heaths (Hampshire/Sussex/Surrey); Dunwich Heath (Suffolk).

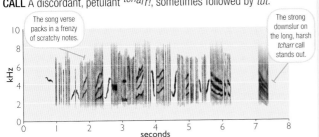

The song verse packs in a frenzy of scratchy notes.

The strong downslur on the long, harsh *tcharr* call stands out.

Firecrest

This gem of a bird looks like a Goldcrest but with bold black and white facial stripes and a red or orangey crown stripe rather than yellow. It bred for the first time in the UK in 1962 and is now well established, males betraying their presence with their simple crescendo song.

SONG Often given high and hidden in trees, it is a beautifully controlled, simple, shimmering trill that gets progressively louder, *si-si-si-si-si-si-si-si-Sl-Sl-Sl-Sl-Sl-Sl-Sl-Sl-Sl*, quite different in structure to the cyclical song of the Goldcrest, although similar in pitch. There are about six to eight notes per second, and the whole song lasts about 2–3 seconds. The trill sometimes accelerates slightly at the start or slows a little at the end.

CALL Its thin *see*, *sip*, *tzi* and *ti* notes are very similar to those of the Goldcrest. They are typically around 8kHz, some rather insistent. It sometimes gives a few *see* notes on a very subtle rising scale.

COMPARE WITH Goldcrest (song and call); Treecreeper (call).

Regulus ignicapilla

UK STATUS Sedentary and increasing resident, possibly now more than 4,000 pairs, mainly in southern England. Also a passage migrant, especially to east and south coasts, and a scarce winter visitor from Europe.

HABITAT Less tied to conifer trees than the Goldcrest but still fond of huge trees in plantations for nesting.

WHERE TO HEAR Try the New Forest or mature plantations in southern England.

Goldcrest

Tracks 26 & 29

This smallest of British birds has ultra high-pitched calls to match, above the hearing range of many people. See pages 69 and 71 for full description.

SONG It starts by repeating a little phrase such as 'sicily' or 'silly-so' three to four times in a cycle, with some crescendo, and finishes with a rather louder, slightly lower-pitched flourish: *si-si-lee si-si-lee si-si-lee sip-sip-sup*. In total, each song lasts about 3–4 seconds.

CALL Makes thin *seeeh* and *si-si-si* calls of various strengths, some as tiny as the bird itself, others of unexpected volume and sibilance, often around 8kHz in pitch but some of the quieter contact calls rather higher. The calls often run into a conversational, extended *sisisisi*....

COMPARE WITH Firecrest (song and call); Treecreeper (song and call); tits (contact calls).

Regulus regulus

UK STATUS Resident, with more than 600,000 pairs, some sedentary, those in upland areas moving down for the winter, plus an influx from northern Europe in autumn.

HABITAT Breeds in coniferous trees. On passage, seen more widely in coastal woods and even scrub.

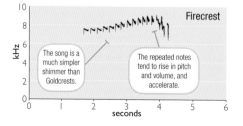

Firecrest

The song is a much simpler shimmer than Goldcrests.

The repeated notes tend to rise in pitch and volume, and accelerate.

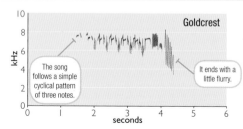

Goldcrest

The song follows a simple cyclical pattern of three notes.

It ends with a little flurry.

Wren

Tracks 18 & 22

Our commonest bird, but you encounter its astonishingly loud song far more than you see it. See pages 62 and 65 for full description.

STAR SOUND: SONG Each vigorous verse is long, typically 4–6 seconds, a rapid-fire hundred notes or more, delivered with gusto and volume. Many notes are very high, giving an overall shrillness. The verse is a run of five to seven trills with short interlinking notes, including a very fast, dry rattle or buzz in the middle of the verse and a rapid-fire trill just before the end.

TIK CALL A metallic *tik*, like marbles knocked together, either singly or in uneven bursts of up to 12 notes per second.

CLOCKWORK CALL A dry rattle like a clockwork toy unwinding, *trrrrrrrt*, given in pulses or short *trits*.

COMPARE WITH Tree Pipit (song); Blackcap (call); Robin (call); Sedge Warbler (call); Redwing (call).

Troglodytes troglodytes

UK STATUS An abundant, sedentary resident with more than 8 million territories spread across almost the entire country.

HABITAT Found in an incredibly wide range of habitats, anywhere where there is dense varied ground cover, from rocky seashores to dense woodland.

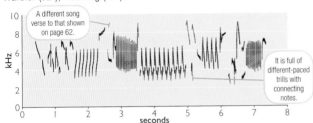

A different song verse to that shown on page 62.

It is full of different-paced trills with connecting notes.

Nuthatch

Track 62

It usually takes a visit to an old woodland to hear its wide and loud repertoire. See page 97 for full description.

'QUIP' CALL Most typically gives a bold, piping *k'wip!*, liquid and pleasant, often in uneven series, and speeding up when excited. In alarm, this becomes a higher-pitched, more urgent '*twip twip…*'.

SONG One version is a fast, piping whinny, lasting 1–2.5 seconds, like a burst of car alarm, easily mistaken for the Lesser Spotted Woodpecker but faster, deeper and purer. Alternatively, gives a slow series of whistles, such as *tyū tyū tyū…* (rather the like the Wood Warbler), *weee weee weee…* or even *wee-oo wee-oo wee-oo….*

OTHER CALLS Understated contact calls include very high-pitched *seet* or *swee* notes, and an equally high, rippling *sirrr*.

COMPARE WITH Willow Tit (song); Wood Warbler (*pyū* song); Lesser Spotted Woodpecker (call).

Sitta europaea

UK STATUS Common and sedentary resident with around 220,000 pairs, widespread in England and Wales, rare but expanding into southern Scotland, and absent from Ireland.

HABITAT Needs mature deciduous trees, whether in woods, parkland or leafy suburbs. It nests in tree holes, adjusting the size of the entrance hole with mud until it is a perfect fit.

Treecreeper

Even though its calls and song are rather subtle and understated, they are nevertheless very useful for alerting you to this tree-trunk-climbing expert. See page 104 for full description.

CALL While climbing up tree trunks, it gives sibilant, high *seeee* notes, rather loud and insistent for calls of such a high pitch. They are longer and with more vibrato than Goldcrest calls and given singly or in slow series compared with the quick bursts of the Long-tailed Tit's calls.

SONG Easily 'overlooked' as Blue Tit or Goldcrest song, this 3-second verse starts with a few slow notes at a very high pitch, but then breaks into a fast descending trill, and ends with a little rising flourish. In contrast, the Blue Tit's trill is straight and without the flourish, and a Goldcrest's verse has a cyclical pattern at the beginning.

COMPARE WITH Goldcrest (song and call); Firecrest (song and call); Long-tailed Tit (call); Blue Tit (song and call).

Certhia familiaris

UK STATUS Rather sedentary and widespread resident, with around 200,000 pairs.

HABITAT Found mainly in larger woods but also in parkland and farmland, where big hedges and riverside corridors connect large trees and copses across the landscape.

Ring Ouzel

This shy thrush of craggy places looks much like a Blackbird but with silvery-edged wings and a pale crescent on the breast. Although sparingly vocal, it has an exquisitely minimalist song and distinctive calls.

SONG The structure is similar to Song Thrush song, each mini-verse typically a note, whistle or brief dextrous phrase that is immediately repeated. There is then a long pause, before a new repeated note or phrase. It has a sparseness, with some notes not repeated at all, or one repeat being enough, or with the repeat notes sounding like an echo.

TCHOK! **CALL** A hard, loud *tchok! tchok!*, often given when flushed, like knocking a stone on an ice-covered pond.

OTHER CALLS Has a wild, laughing rattle, rather Fieldfare-like, *schak-schak-schak-…*, more musical than the *chok* call. May also give a rapid vibrato *brrrrp*.

COMPARE WITH Song Thrush (song); Blackbird (call); Fieldfare (call).

Turdus torquatus

UK STATUS Scarce summer visitor, breeding mainly in north Wales, northern England and Scotland, with around 7,000 pairs. On passage, mostly in south and east England.

HABITAT Breeds in upland areas with rocky crags, steep gullies and abandoned mines, feeding on moorland and upland pasture; on passage, visits downland and coastal scrub.

WHERE TO HEAR Breeding sites include: Snowdonia (Gwynedd); Stanage Edge (Derbyshire/South Yorkshire); Rosedale (North York Moors); and Cairn Gorm (Highlands). In autumn, try Beachy Head (East Sussex).

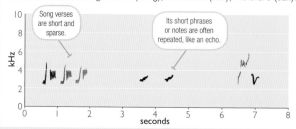

Song verses are short and sparse.

Its short phrases or notes are often repeated, like an echo.

Starling

Track 40

Still a feature in many gardens, despite large declines, this star-covered bird is not only feisty and characterful but also has an incredible song and plenty of other calls.

STAR SOUND: SONG See page 80 for full description. Each extended verse, up to a minute long, is a fairly quiet jumble of electric squiggles, blips, pops and chuckles, often including long, whizzing whistles and all sorts of imitations.

'STARLING CHORUS' Flocks will gather to sing together, often for long periods. They sometimes fall silent simultaneously, which is known as a 'dread'.

CALL A vibrato *schree* or *scheer*, especially when taking flight, heard endlessly from family parties in summer.

ALARM CALL A long, dry, rasping *tcharrrr*, like a small Jay, and often makes 'chackering' noises when bickering over food. Also gives a brief *stek stek* when announcing that it has spotted a falcon or hawk.

COMPARE WITH Jay (call).

Sturnus vulgaris

UK STATUS Abundant but much declined resident across much of Britain, with around 2 million pairs, boosted in winter by large numbers of continental birds.

HABITAT In farmland, feeds on short grassland; in urban areas, visits parks, gardens and sports pitches. It is also found on seaweed-strewn beaches and saltmarsh.

Blackbird

Tracks 15 & 21

Such a familiar garden bird, knowing its beautiful song and wide repertoire of other calls is the bedrock to learning other bird calls. See pages 60 and 64 for full description.

STAR SOUND: SONG Ever-changing, beautiful, relaxed verses, each 2–4 seconds, starting with five to seven complex, confident, mellow notes, and ending with a short, anticlimactic, often high-pitched twiddle. Each male may have 30 or more different verses.

***WIK* CALL** A staple sound at dusk, or when scolding a cat or owl, it is an incessant, loud, sharp *wik wik wik...* in long, uneven series.

***CHOK* CALL AND PANIC RATTLE** Given in short series, the deep *chok chok chok...* sometimes breaks out in a total panic before calming down again.

OTHER CALLS A very thin, long *tseeeeee* warns of an aerial predator. Also has a deep, quiet *hic* of mild anxiety in slow, uneven series, and migrants give a short, high, sibilant *seee*.

COMPARE WITH Mistle Thrush (song); Robin (song); Song Thrush (calls); Redwing (calls).

Turdus merula

UK STATUS Abundant resident breeder across the country, with around 5 million pairs, plus passage migrant and winter visitor from northern Europe and the near continent.

HABITAT Gardens, parks and woods, wherever there are trees and scrub plus bare or short-turfed ground.

Fieldfare

This fine-looking, robust thrush provides one of the defining sounds of winter with its loud *tchak-tchak* calls as large flocks gather in the countryside, in harsh weather even pushing into gardens.

***TCHAK* CALL** This striking call is given singly or often in doubles, triples or sometimes more, at times with a hoarse, wild laughing quality, and sometimes introduced with a quieter *po*. It also gives an even drier *trrak-trrak*, like a couple of bursts of a camera shutter.

***WIH* CALL** A high, weak, rather breathy *wih*.

SONG Rarely heard in Britain, this anonymous twiddling twitter is either in short 1-second verses, or longer and more ecstatic in song flight. Its verbal scribbles are like an extended ending of Blackbird song, with plenty of high pitch but interspersed with *trak-trak* calls.

COMPARE WITH Mistle Thrush (call); Ring Ouzel (call).

Turdus pilaris

UK STATUS Nomadic widespread winter visitor from Scandinavia, present mainly October to March although a few linger to May, with around 700,000 individuals. Just one or two pairs stay to breed in Scotland and northern England.

HABITAT Mainly farmland, feeding out on open short turf and also in tall hedgerows and orchards.

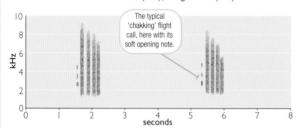

The typical 'chakking' flight call, here with its soft opening note.

Song Thrush

The Song Thrush lives up to its name, giving one of the most inventive and distinctive songs, a favourite for many people.

STAR SOUND: SONG See page 88 for full description. The song is rich and loud, with an unusual 'short phrase, repeat, move on' structure. Each phrase is short, sometimes no more than a single note, and the number of repeats varies between one and five times. Every once in a while, it will sing a more complex, fiddly phrase that it does not repeat. The Song Thrush is often a lead voice in the dusk chorus and can sing for long periods, making this a stand-out, inventive and arresting vocalist.

***STIP* CALL** Often heard as birds take off or when migrating overhead, it is a high-pitched, abrupt *stip* or *tip*, so high it seems you half hear it.

ALARM CALL A *chuk chuk* or *chop chop*, often in twos or threes, and a fast, higher *chik-ik-ik-ik-ik-ik*, very similar to the Blackbird's alarm.

COMPARE WITH Ring Ouzel (song); Blackbird (call).

Turdus philomelos

UK STATUS Abundant but much declined resident across almost all of the country, with about 1.2 million pairs. It is also a passage migrant, and a winter visitor in large numbers.

HABITAT Parks, gardens, deciduous woodland, and mixed farmland with thick hedges, ditches and damp pasture.

Redwing

This winter thrush has a surprisingly wide range of subtle calls, including the main one to listen for on autumn nights.

***TSEEEE* CALL** A thin yet strong and sibilant *tseee* or *tssss*, slurring down from an ultra-high start, given by night migrants overhead but also by day. Beware the similar call of the Blackbird.

***HIK* CALL** An anxiety call, rather quiet, short and deep in pitch, sounding like a slightly tipsy hiccup.

***TRIT* ALARM CALL** Very similar to the *trit* call of the Wren, with rolling 'r's.

SONG Very rarely heard in the UK, each verse starts with a simple, melodic series of about five confident whistles, running up or down the scale, before switching to a formless, introspective twitter, up to 30 seconds long. On fine early spring days, flocks in hedges and trees may twitter quietly together.

COMPARE WITH Song Thrush (call); Wren (call); Blackbird (call).

Turdus iliacus

UK STATUS Winter visitor from northern Europe across much of the UK, with around 700,000 individuals, mainly from October to March but a few lingering until May. Fewer than 10 pairs breed in northern Scotland.

HABITAT Largely found in mixed farmland with grassy pastures, berry-rich hedgerows and woods, moving into gardens and parks in hard weather.

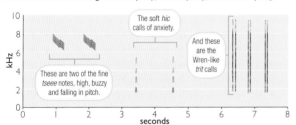

The soft *hic* calls of anxiety.

And these are the Wren-like *trit* calls

These are two of the fine *tseee* notes, high, buzzy and falling in pitch.

Mistle Thrush

Tracks 51 & 52

The melancholy song and 'football rattle' call help separate this large thrush from the Song Thrush. See page 89 for full description.

STAR SOUND: SONG Each short verse is like pared-back Blackbird song with just three to six simpler fluty notes, like rhyming lines of a wistful poem. There is often a short pause of a couple of seconds between verses, and it can often sing for long periods at a time. Its song is usually given from a very high perch in a tree, and is notable for continuing even in poor weather.

'FOOTBALL RATTLE' CALL A long, wooden *prrrrrrrrrrt*, in alarm broken into a series of short rattles, *trrt trrt trrt trrt trrt….*

COMPARE WITH Blackbird (song), Magpie (call).

Turdus viscivorus

UK STATUS Fairly common but declining resident, with about 170,000 pairs.

HABITAT Feeds on open dry grasslands but sings and nests in tall trees, so found in parkland and heathland and especially open upland forests. In winter it guards trees laden with berries.

Spotted Flycatcher

Track 72

Although unaccomplished vocally, the little disjointed calls and paltry song can nevertheless help locate this bird high in the trees. See page 107 for full description.

CALL High ^{tsi} notes can be given singly or, when the bird is more alarmed, are followed by lower *chuk* sounds, such as ^{tsi} *chuk-chuk* or ^{tsi} *chu-chu-chuk.*

SONG The rudimentary verses are little more than an embellished series of disconnected call notes, at an unsteady rhythm of around one per second, such as ^{tsi} *tsup dizi zwi zit-chuk dzi*.... It is a quiet, reserved song, and is only given occasionally. As a result, it is very easy to overlook among the more attention-grabbing woodland songs.

COMPARE WITH Redstart (call); Robin (call); Goldcrest (call); Treecreeper (call); tit species (calls).

Muscicapa striata

UK STATUS Widespread but much declined summer visitor and passage migrant, with probably no more than 36,000 pairs.

HABITAT Now most familiar in northern and western ancient woodlands, especially where there are open glades or along streamsides.

Robin

Track 16 & 2

The Robin's ever-changing song verses are watery perfection, heard almost entirely year-round, plus it has a range of calls. See pages 61 and 64 for full description.

STAR SOUND: SONG The short verses, typically lasting just 1–3 seconds but sometimes more, are beautiful and laid-back, shifting between high- and low-pitched phrases. They are extremely variable, but have a consistent watery feel, gurgling and trickling like a mountain stream, with long notes interspersed with rippling cascades. The territorial song by both sexes in autumn and winter has a more wistful tone.

TIK CALL A rather sharp *tik* or *tip*, either singly or in stuttering series.

ALARM CALL A strong, piercing ^{seeee}, slowly repeated, very high in pitch.

CONTACT CALL A quiet *swi*, brief and rather occasional.

COMPARE WITH Blackbird (song); Blackcap (call); Dunnock (song); Spotted Flycatcher (call).

Erithacus rubecula

UK STATUS Abundant, widespread and mainly sedentary resident with more than 6.5 million breeding pairs. It is also a passage migrant in autumn from northern Europe.

HABITAT Mainly woodland, scrub, parks and gardens, where there are elevated song perches plus plenty of cover and cool, moist shade.

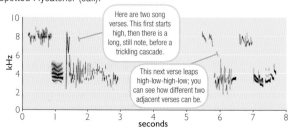

Here are two song verses. This first starts high, then there is a long, still note, before a trickling cascade.

This next verse leaps high-low-high-low; you can see how different two adjacent verses can be.

Nightingale

Our most famous singer combines power, creativity and ultimate vocal control in a standout sound that is worthy of its reputation.

STAR SOUND: SONG See page 95 for full description. Most verses last about 2–3 seconds and have a twiddly introduction, then a couple of rapid-fire series, and a final acrobatic note or two. However, a few begin *tyooo tyooo tyooo…*, slowly accelerating over many seconds into an explosive ending. The song seems incredibly assured and constantly inventive, and there is a boldness, clarity and volume to the sound which is its hallmark.

FROG-CROAK CALL A deep, dry croak, *rrrrrp*, like running a stick down a notched block of wood. It is rather quiet and unobtrusive.

SWEE CALL A simple *swee*, similar to the Chiffchaff but shorter.

COMPARE WITH Chiffchaff (call); Redstart (call).

Luscinia megarhynchos

UK STATUS Scarce summer visitor, with fewer than 5,500 singing males remaining, now restricted to mainly south-east England and East Anglia. It arrives from mid-April and most leave in August.

HABITAT Once a bird of coppice woodland, it is now more often found in scrubby areas near water, with bare ground beneath.

WHERE TO HEAR Pulborough Brooks (West Sussex); Highnam Woods (Gloucestershire); Northward Hill and Blean Woods (Kent); Fingringhoe Wick (Essex); Minsmere (Suffolk).

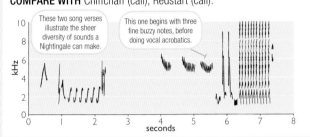

These two song verses illustrate the sheer diversity of sounds a Nightingale can make.

This one begins with three fine buzzy notes, before doing vocal acrobatics.

Pied Flycatcher

For just a few weeks, from late April to early June, the male's simple, song is part of the steady accompaniment to the distinctive spring chorus in upland oakwoods.

SONG See page 109 for full description. Each bright verse uses just a few basic notes spread over a limited pitch range, strung together randomly for about 5–10 notes. The pacing is fairly consistent, to a pedestrian plodding beat, with a few notes repeated, conjuring an image of someone doing step exercises.

CALL The wide range of short calls includes a sharp little whistle, *swik*, plus a harder, deeper, metallic *tuk*, used in alarm. It often combines the two: *swi-tuc*.

COMPARE WITH Redstart (song); Black Redstart (call); Whinchat (call); Spotted Flycatcher (call).

Ficedula hypoleuca

UK STATUS Summer visitor to western areas, with fewer than 20,000 pairs; very rare in Ireland. Also a scarce passage migrant along east and south coasts in spring and especially autumn.

HABITAT Mainly ancient oak woodlands in upland valleys, nesting in tree holes but very quick to take advantage of nestboxes.

Black Redstart

Looking like a Redstart dunked in soot but with its quivering tail still glowing red, the male has a curious fragmented song that can be vital for locating males in their urban 'wastelands'.

SONG Perched up on a roof, the male starts each verse with a short, rapid series of four to six sharp, bouncing notes. He then pauses for up to 2 seconds, then gives a quiet, vague, crunching sound, like scrunching a handful of gravel, before giving a different bouncing series to end. The first series often rises subtly in pitch and the second is often lower in pitch, so one song verse is as if the male asks a question, his brain rumbles and he then gives the answer. At a distance, it is easy to miss the gravel noises and indeed he will sometimes sing just one or the other part of the verse.

CALL The two typical calls are a weak, high-pitched *si* whistle and a harder *trit* or *tek*; as with the other chats and the Redstart, these are often run together: *si tek   si te-tek....*

Phoenicurus ochruros

UK STATUS Rare breeding visitor with about 20–40 pairs, mainly in south and east England; also a passage migrant and winter visitor, mostly in southern coastal areas, with up to 400 present in midwinter.

HABITAT In the breeding season, power stations, docks and old industrial areas. In winter, typically coastal, again often in industrial areas or around old forts or undercliffs.

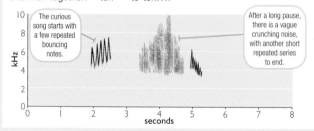

The curious song starts with a few repeated bouncing notes.

After a long pause, there is a vague crunching noise, with another short repeated series to end.

Redstart

Track 76

This is one of the more anonymous songs of upland woodlands, each short verse starting with confidence but losing its way.

SONG See page 110 for full description. The unspectacular song verse starts with a plaintive *soo*, followed by a confident, fast series of two to five bouncing notes, rather like the Chaffinch but then running out of steam, disintegrating into a variable short warble. So the key aide-memoire is 'same confident start, different weak endings'.

CALL There are two common calls: a plain, upslurred *hwee*, very similar to that of the Chiffchaff and the Chaffinch, and a much harder *tip* or *tik* call, often alternating between the two.

COMPARE WITH Chaffinch (song and call); Black Redstart (song and call); Willow Warbler (call); Chiffchaff (call); Spotted Flycatcher (call); Nightingale (call).

Phoenicurus phoenicurus

UK STATUS Summer visitor with around 100,000 pairs, most in Wales, northern England and Scotland. Passage migrants are seen mainly near the coast and especially in autumn.

HABITAT Breeds in ancient oak woodland, open native pinewoods and parkland with mature trees, nesting among tree roots and along banks and walls.

Whinchat

The song of this declining chat is very varied, and is difficult to describe and learn, but is no less interesting for that.

SONG Each brief, 1-second verse is made up of two to four distinct phrases. These are hugely varied, including plenty of sweet twitters, but also scratchy or squelchy notes, strange crunching and crackling noises as if walking on gravel or scrunching up paper, plus snatches of imitation of birds including the Corn Bunting's jangling keys. So one verse might be: 'Robin imitation–*zweet-zweet*–random twitter', the next 'fizzy sound–deep notes–*zwit-zwit*'. It is wonderfully odd!

CALL Its main hard chat-note is a high-pitched *tik*, sometimes two or three run quickly together, often used in combination with a sad, whistled note, *Soo*, much like the Bullfinch call. The overall effect is distinctive: *Soo tik Soo t'tik....*

COMPARE WITH Wheatear (song and call); Stonechat (song and call).

Saxicola rubetra

UK STATUS Summer visitor, now largely restricted to uplands in Wales, northern England and Scotland, with fewer than 50,000 pairs remaining. More widespread but scarce on passage, especially on south and east coasts.

HABITAT Now mainly found in upland valleys, where bracken and scattered bushes hug damp, grassy slopes.

Wheatear

Mostly silent on migration, it is on its mountain breeding grounds that the Wheatear gives a song that almost defies description.

SONG Usually delivered from a prominent rock, each verse is typically a mishmash of different, garbled sounds crammed into short 1–1.5-second bursts, mixing sweet notes with deep crunching noises and sounds like a camera shutter. It often makes two very different sounds simultaneously! It sometimes includes impressive short imitations of other birds, and does an extended version during song flights. Verses are sometimes repeated many times, sometimes not. Its variability means telling it from the song of the Whinchat can be difficult.

ZIP TUK **CALL** It alternates sweet and hard notes, in the case of the Wheatear a deep *tuk* and a higher, soft *Zip*.

COMPARE WITH Whinchat (song and call); Stonechat (call).

Oenanthe oenanthe

UK STATUS Common summer visitor across much of upland UK and some coastal areas, with almost 250,000 pairs. Also a passage migrant.

HABITAT On migration, seeks open habitats such as short turf; in the breeding season, mainly in open uplands.

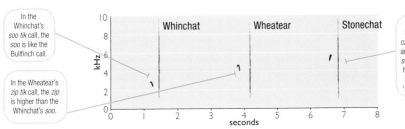

In the Whinchat's *soo tik* call, the *soo* is like the Bullfinch call.

In the Wheatear's *zip tik* call, the *zip* is higher than the Whinchat's *soo*.

Whinchat   Wheatear   Stonechat

And the Stonechat's call starts with an even higher *swee*; with the harsh note, it sounds like 'sweet shop'.

Stonechat

Often seen sitting on gorse bushes, this is the sentinel of heathy places.

SONG Each short, anonymous verse is only 1–1.5 seconds long, just six to eight rather scratchy notes, and is repeated a few times before moving to the next. It all feels 'samey', an anonymous verbal scribble like a snatch of Whitethroat song, especially in occasional fluttering song flight, but without the up-and-down, 'sing-song' feel.

'SWEET SHOP' CALL A whistled *swee* alternating with a rather deep and hard *chrt*, giving what can be remembered as 'sweet shop'.

COMPARE WITH Whinchat (song and call); Wheatear (song and call); Dunnock (song); Whitethroat (song).

Saxicola rubicola

UK STATUS Fairly common resident with around 60,000 pairs. Some are sedentary, but others, especially those that breed on higher ground, are short- or medium-range migrants, moving to milder areas for the winter.

HABITAT Most associated with dry, warm, rough, open habitats with short scrub, such as lowland and coastal heathland.

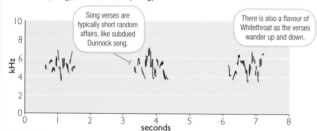

Song verses are typically short random affairs, like subdued Dunnock song.

There is also a flavour of Whitethroat as the verses wander up and down.

Dipper

Learning the penetrating flight call will help you find this amazing daredevil swimmer along its upland streams.

CALL A short, hard, rather metallic *tzik* or *tcheek*, loud enough to be heard above the noise of a mountain stream.

SONG Both sexes sing, each verse a steady chatter of sparse random notes, about four per second and often for 10–30 seconds. It resembles the pace and structure of Reed Warbler song, but made up of electronic squeaks and whistles, high *zip* notes and longer notes that sound like slurps or squelching through mud. It feels like a song playing backwards.

COMPARE WITH Kingfisher (call).

Cinclus cinclus

UK STATUS Rather scarce, sedentary resident in the west and north, with around 11,000 pairs.

HABITAT Tied almost exclusively to clean, fast-flowing, shallow and rocky streams and rivers, commonest in chalk and limestone areas and characteristic of upland valleys.

WHERE TO HEAR Great sites include: Dovedale (Derbyshire); Bolton Abbey (North Yorkshire); the Falls of Clyde (South Lanarkshire).

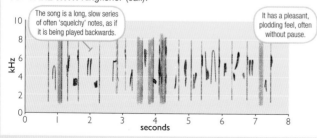

The song is a long, slow series of often 'squelchy' notes, as if it is being played backwards.

It has a pleasant, plodding feel, often without pause.

House Sparrow

One of the nation's best-known birds, it is hardly musical but is definitely cheerful and chatty! See page 79 for full description.

CALL The typical call is variations on a single chirping note, $ch^{il'}p$ – bright, short and confident with a 'half-jump' in the middle.

SONG Each 'verse' is just three or four variations on its call note, in uneven series at just one to two notes per second: $ch^{il}p$ $ch^{ow}p$ $sh^{ee}lp$ $ch^{il}p\ldots$, etc.

THREAT CALL When feeding or jostling for a perch, it gives uneven, tetchy rattles, some short, some longer, tending to wander in pitch a little.

'SPARROWS CHAPEL' Flocks of sparrows gather in favourite bushes to chat noisily, a mix of main calls and soft, deep, muttered *churp* and *cheep* sounds.

COMPARE WITH Tree Sparrow (song and call); Reed Bunting (song).

Passer domesticus

UK STATUS Sedentary and very widespread resident, with around 5 million pairs; absent from the highest ground.

HABITAT Almost always associated with farms, houses and gardens, rural and urban, nesting in holes in buildings.

Tree Sparrow

The scarce country cousin of the House Sparrow, with a chestnut cap and black patch in the middle of its white cheeks, is very similar in its calls and song but with a few subtle differences to help you.

CALL As conversational as the House Sparrow, it shares the core *tchilp* note. However, it has a slightly wider range of calls, including more subtly musical notes, plus some that are higher and others lower pitched, including variations on *tchŭp* and *tcherp*, which have a bit more of a toy trumpet feel. Many notes are also shorter and more clipped, such as *chip* and *tik* or *ti-tik* calls, especially given in flight. Also listen for a higher-pitched *tch'witt* call.

THREAT CALL A rapid *tet-tet-tet-tet-tet-tet*..., similar to the House Sparrow's *chir'r'r'r'p* but more wooden.

SONG Rudimentary, no more than a slow series of its call notes.

COMPARE WITH House Sparrow (songs and call); Wren (call).

Passer montanus

UK STATUS Much declined sedentary resident, with around 200,000 territories, mainly from the English Midlands northwards, with a coastal and eastern bias. The small Irish population is growing fast.

HABITAT Mainly farmland, nesting in tree holes or nestboxes in tall hedgerows, small copses or willows along ditches and streams, and feeding in weedy stubbles in winter. Rare in urban areas.

WHERE TO HEAR Old Moor (South Yorkshire); Blacktoft Sands and Bempton Cliffs (East Yorkshire); Fairburn Ings (West Yorkshire); Loch of Strathbeg (Aberdeenshire); Portmore Lough (Co. Antrim).

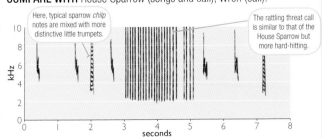

Here, typical sparrow *chilp* notes are mixed with more distinctive little trumpets.

The rattling threat call is similar to that of the House Sparrow but more hard-hitting.

Dunnock

Both the bird and its hurried little ditty are rather modest and seldom recognised, despite being so widespread in gardens and other scrubby habitats. See pages 63 and 65 for full description.

SONG Each verse is a rather long, fast, squeaky ditty, usually lasting 2–4 seconds. It shuttles along at about eight notes per second like a line of verbal scribble, meandering up and down in a samey way without pause or change in pace. It doesn't have the Robin's gurgling notes or changes of pace, nor the harsh notes of the Whitethroat.

TZEE CALL A simple *tzee* or just *tzzz*, rather weak and sibilant, like a squeak of the hinges of a rusty gate. It is often given in doubles or as a little trill of four to six notes. Pairs and trios often respond to each other.

COMPARE WITH Robin (song); Whitethroat (song); Kingfisher (call); Stonechat (song); Dartford Warbler (song).

Prunella modularis

UK STATUS Sedentary and widespread resident with around 2.3 million territories, rare only in mountain areas and in the far north and west.

HABITAT Found wherever scrubby bushes grow densely, such as in gardens, town parks and hedgerows, avoiding either very open or well-wooded habitats.

Yellow Wagtail

The best way to find this colourful summer visitor is either to look for small groups skipping around the feet of grazing cattle or to learn the diagnostic flight call.

SWEET **CALL** Not especially vocal, calling mostly in flight, an uncomplicated, bright *sweet!*, often given singly or, if repeated, at a rate of about one per second. If heard clearly, you sense two half-syllables run quickly together, shooting up to very high pitch and instantly dropping again, *swe-ut*, giving each call a 'finished' feel.

SONG Paltry – little more than elaborations of the calls, often in slow series while sitting on the ground or a wire fence, with some notes in quick pairs. A few notes are longer and more rippling, *psir'r'r'rp*, slurring strongly down.

Motacilla flava

UK STATUS Declining summer visitor and passage migrant, with 15,000 pairs remaining, most in the Midlands and East Anglian Fens through to Yorkshire, especially in the east. Passage migrants are mostly seen along east and south coasts. Absent from Ireland.

HABITAT Flat, extensive wet pasture, grazing marshes and oil-seed rape fields.

WHERE TO HEAR Ouse Washes (Cambridgeshire/Norfolk); Lower Derwent Valley (East Yorkshire).

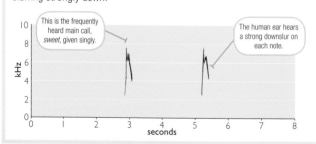

This is the frequently heard main call, *sweet*, given singly.

The human ear hears a strong downslur on each note.

Grey Wagtail

This signature long-tailed bird of fast-flowing streams, weirs and cascades needs to be heard over the noise of the water, so has a piercing call and song.

CALL See page 84 for full description. Similar to the Pied Wagtail's *chissik*, this double-note *zit-zit* or *zi-zit* is sharper and more penetrating, with identical notes.

SONG Sings more frequently than the Pied Wagtail, each verse a short, straight, fast series of identical notes, higher than the call but with the same piercing, mechanical quality, such as *szit-szit-szit-szit-szit*, or an even faster *siz-iz-iz-iz-iz*.

ANXIETY CALL An insistent *see^{ep}* or *sū-e^{eep}*, rising strongly, often given repeatedly interspersed with the main call note.

COMPARE WITH Pied Wagtail (call); Common Sandpiper (call).

Motacilla cinerea

UK STATUS Resident, with around 38,000 pairs, spread widely but most abundant in upland areas. Some are sedentary, but many upland birds move south and west for the winter.

HABITAT Is attracted to fast-flowing and gushing water with surrounding trees, so is common along upland streams but is also drawn to weirs and mill races. It is more widespread on slower waters in winter.

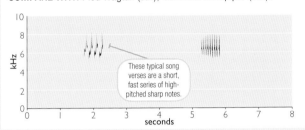

These typical song verses are a short, fast series of high-pitched sharp notes.

Pied Wagtail

The 'car park bird' is our own little black-and-white roadrunner, with a bright flight call best remembered as the 'Chiswick flyover' but with barely a song to its name.

FLIGHT CALL See page 83 for full description. Often given on take-off, it is a sharp *chi^{z}zik*, usually two syllables but sometimes one or three. The two syllables clearly different with the second often slightly lower in pitch, in contrast to the two identical notes of the otherwise similar Grey Wagtail call.

TERRITORIAL CALL More often given on the ground or a perch, it is also usually two syllables but a more musical, slurred *swiz-z^{ee}*, *shwer-z^{ee}* or *che-w^{ee}*. Similar sweet notes are also given when entering roosts.

SONG Rarely heard, it either gives little more than a series of slightly elaborate territorial calls, or very occasionally loses itself in a faster, excited, uncontrolled version, interspersed with deeper sweet notes.

COMPARE WITH Grey Wagtail (call).

Motacilla alba

UK STATUS The Pied Wagtail is the British race, a widespread resident, with nearly 500,000 pairs. The White Wagtail is the continental race, a spring and autumn migrant.

HABITAT Bare ground and very short vegetation, including seaweed beaches, ploughed fields and sewage farms. Nests in barns and stone walls. In winter, flocks roost in reedbeds, city centre trees and warehouse roofs.

Meadow Pipit

Track
88

This is the essential voice to accompany you on moorland walks in spring and summer. See page 121 for full description.

SONG Starts with a controlled *tsip-tsip-tsip-...* or *tew-tew-tew-...*, the notes slightly sharp, at about five per second. Many verses are false starts, but in full song, often in song flight, the opening notes crescendo and then break into a chain of different trills, some chinking, some rattling, lasting 20 seconds or more.

CALL Frequently given, especially on take-off, it is a rather sweet, pitiful *sip* or *seep*, high-pitched, slightly impure and lisping. Telling it from the Rock Pipit is very difficult, as it sometimes almost has the fizz of its cousin's *feest!* At times, the Meadow Pipit extends its calls into a fast series, falling sequentially in pitch, *sip-sip-sip-sip-sip-sip*.

ALARM CALL Two weak notes in quick succession, sounding to our ears as *t'rip*, repeated incessantly.

COMPARE WITH Rock Pipit (song and call); Tree Pipit (song).

Anthus pratensis

UK STATUS Abundant, widespread resident and summer visitor, with around 2 million pairs, the largest populations in the north and west. Also a passage migrant and winter visitor from Scandinavia and Iceland.

HABITAT Open, semi-natural habitats, including downland, saltmarsh and wet grassland but especially on upland moors in summer. Avoids enclosed places such as woods and gardens.

Tree Pipit

Track
77

Both the song and the call of this upland and heathland specialist are critical for identification.

STAR SOUND: SONG See page 111 for full description. Rapturous, each verse has two to four clear phrases that are linked seamlessly, each a second or more long and including fast trills, seesawing couplets and whizzing repeated notes. In song flight, the male adds more series, extending the song to 12 seconds or more, and often using repeated long *weeeee* notes towards the end as if he really is having fun!

CALL A distinctive, fizzing, high-pitched *spiz*, at about 6.5kHz, given singly or at most a couple of times a second. It is like a buzzy Dunnock call, and is a key way of identifying migrating birds flying over.

COMPARE WITH Wren (song); Meadow Pipit (song); Dunnock (call).

Anthus trivialis

UK STATUS Summer visitor, arriving from late March, with perhaps 90,000 pairs, mainly in the west and north. Also a passage migrant, mainly at coastal sites.

HABITAT A bird of open woodlands, woodland edge, young plantations and native pine woodlands in the uplands, but also found on lowland heathlands and commons with scattered trees. It feeds mainly on the ground.

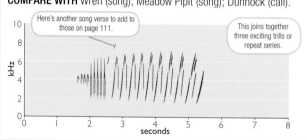

Here's another song verse to add to those on page 111.

This joins together three exciting trills or repeat series.

Rock Pipit and Water Pipit

The Rock Pipit (right) looks like a sooty Meadow Pipit, with a song and calls that need to be heard over the sound of the sea in its rocky, coastal haunts. The Water Pipit is a rare winter visitor with an identical repertoire.

SONG In structure, very similar to the Meadow Pipit. An initial series of sharp notes, *jig-jig-jig…*, runs for several seconds, frequently petering out. However, in song flight the opening series accelerates and ultimately breaks into a chain of two to five different series, each at a different tempo, including some notes that sweep up in pitch like in Tree Pipit song. The notes tend to be harder and more 'chinking' than in the Meadow Pipit, and the hard 'g' ending to the *jig-jig* notes is characteristic.

CALL A rather electric *feest!* or *wheesp!*, fuzzier and more forceful than most Meadow Pipit calls, although weaker calls are hard to tell apart. The rising pitch gives the call an urgent feel. It is almost always given singly.

COMPARE WITH Meadow Pipit (song and call).

Anthus petrosus
Anthus spinoletta

UK STATUS The Rock Pipit is a sedentary resident, with around 35,000 pairs; in addition, some birds from Scandinavia arrive for the winter. Just 200 Water Pipits from Europe winter here.

HABITAT In the breeding season, the Rock Pipit is almost exclusively tied to rocky coasts with cliffs; in winter, some birds also use open saltmarsh. Our wintering Water Pipits tend to visit boggy meadows and freshwater pool edges.

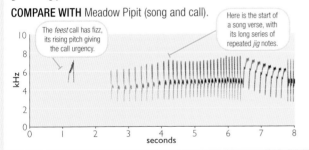

The *feest* call has fizz, its rising pitch giving the call urgency.

Here is the start of a song verse, with its long series of repeated *jig* notes.

Richard's Pipit

This rare, long-legged pipit is large compared to our Meadow Pipit, but has a habit of feeding deep amongst long grasses and so spends much of its time hidden from view. Knowing its strident call is incredibly useful for picking it out.

CALLS The main call, given especially when rising from the ground but also regularly in flight, is a bold and loud *shreep!*, with the transcription being accurate to most ears. It is short and punchy, with an assertive rise and fall in pitch, and is considerably deeper in pitch than the plaintive *seep* of a Meadow Pipit. Overall, it is surprisingly similar to the call of a House Sparrow. It shows little variation, so remembering 'sparrow shreep' will help fix it in your mind.

COMPARE WITH Skylark (call).

Anthus richardi

UK STATUS Rare autumn visitor from Siberia, with around 100 seen each year, mainly along the east coast from Shetland to Kent, and in Cornwall and Scilly. A very few winter and are seen in spring.

HABITAT Open coastal grasslands, especially where the grass has been left uncut.

Chaffinch

One of our commonest birds, with a wide repertoire of key sounds, including the 'bold jig' song. See pages 75 and 76 for full description.

SONG Neat, bright and brash, each 2–3-second verse skips down the scale, at each step giving a short, quick series of two to six repeated notes, and at the bottom rounds off with a flourish, *cheweeoo*, often rising. A male has up to six verses in his repertoire, repeating one many times before moving to the next.

***WINK!* CALL** A strong, bold, clear *wink!* or *chink!*, often repeated and difficult to tell from the Great Tit's cheerful *pink!*

'RAIN CALL' In spring and summer, males utter one note repeatedly, tediously, about once a second, for many minutes, such as *hweet*, *zreep* and *wi-jit*.

FLIGHT CALL A soft *tyūp*.

COMPARE WITH Redstart (song and call); Willow Warbler (song); Great Tit (call); Chiffchaff (call), Brambling (call).

Fringilla coelebs

UK STATUS Abundant and very widespread resident, with more than 6 million territories, bolstered by large numbers of passage migrants and winter visitors from northern Europe.

HABITAT Found wherever there are trees and bushes, especially woodlands but also parks and larger gardens. In winter, a common visitor to gardens and arable stubbles.

Brambling

The nasal call, interspersed with its clipped *chup!* notes, will help you pick out this dapper winter finch amongst Chaffinch flocks.

NASAL CALL Calls readily in flight, a distinctive *che-wee*, incredibly nasal and standing out clearly from other common finch calls. It is deep, with a clear warped rise in pitch, almost with the feel of a little parrot!

OTHER CALLS In flight, gives a clipped *chup!* note, harder than the Chaffinch's *tyup*. Also has a sharper, loud alarm call, *tchik!*

SONG Occasionally heard from migrants in April, it is just a long, deep, fuzzy buzz, *zweeeee*. It is very similar to the *dweeeeee* of the Greenfinch but is flat in pitch with just a hint of a waver, and it starts weakly, increasing in volume and ending rather abruptly.

COMPARE WITH Greenfinch (song); Chaffinch (calls).

Fringilla montifringilla

UK STATUS Passage migrant and winter visitor from northern Europe, present October–April in extremely variable numbers, with up to 1.8 million estimated in some winters, but perhaps only 50,000 in others. It is a very rare breeder, with occasionally one or two pairs in Scotland.

HABITAT In winter, found especially in beech forests where flocks feed on beechmast. Also visits deciduous woods, farmland and some garden feeders, often in the company of Chaffinches.

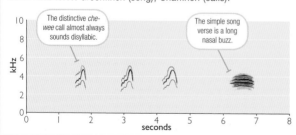

The distinctive *che-wee* call almost always sounds disyllabic.

The simple song verse is a long nasal buzz.

Hawfinch

All too often, this magnificent finch is seen only in chunky silhouette at the top of a tree, or glimpsed as it bolts up from the woodland floor. It is not at all vocal but does have a key call to help you find it.

TZIK! CALL Distinctive but still easy to miss, it is similar to the *tik* call of the Robin but it has a whipped quality, like two flints glancing against each other, almost *stik!*

TSEE CALL Unexpectedly high-pitched for such a large finch, this thin call is easy to overlook.

SONG Rarely heard and frankly rather pathetic, it just uses the two main calls and another couple of heavily clipped sounds in a sparse series, such as *tik! tsee tū tik! tik!....* At the song's 'peak', it may throw in occasional slightly longer, half-strangled, squeaky bicycle-wheel notes, and increase the tempo a little.

COMPARE WITH Robin (call); Redwing (call); Bullfinch (song).

Coccothraustes coccothraustes

UK STATUS Scarce and declining resident, widespread but perhaps just 500–1,000 pairs; absent from Ireland. Also winter visitor in small numbers from the continent.

HABITAT Mature and extensive forest, especially associated with Hornbeam and Yew trees in winter. Often loyal to traditional roost sites in winter.

WHERE TO HEAR New Forest (Hampshire); Lynford Arboretum (Norfolk); Parkend, Forest of Dean (Gloucestershire); Clumber Park (Nottinghamshire); Sizergh Castle (Cumbria); Scone Palace (Perth and Kinross).

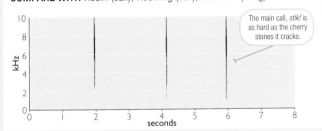

The main call, *stik!* is as hard as the cherry stones it cracks.

Bullfinch

Track 70

If there is one bird that sounds meek and embarrassed, it is this one, but its soft calls are important for locating it in dense hedges.

PHŪ CALL See page 105 for full description. Pitiful, half-hearted and muffled, this main call is used by a pair to stay in contact. Each note is rather deep in pitch, with a slight downslur.

WUT CALL This call is very brief and even quieter and deeper than the main call, usually given just before flight.

SQUEAKY BICYCLE SONG Rarely heard, each very quiet verse is a string of sparse and simple notes at a rate of about one a second. The notes are quite varied, some like extended call notes, others warped or wheezing, with little repetition; it sounds rather like a distant squeaky bicycle being wheeled slowly along. Some of the notes contain simultaneous whistles on two pitches, one rising, one falling, sounding 'off-key' as a result.

Pyrrhula pyrrhula

UK STATUS Sedentary resident, with more than 200,000 pairs, widely if thinly distributed everywhere except the islands of Scotland, with a few migrants arriving from northern Europe in autumn.

HABITAT Woodland, scrub, mature hedgerows and larger gardens, in spring moving into orchards to eat fruit blossom.

Greenfinch

Hard hit by disease, this is nevertheless still a common garden bird with a gloriously smooth voice. See page 78 for full description.

CALLS Has a range of attractive calls, the most frequent being a little trill of rather sweet ringing notes, *ji-ji-ji-*... or *dibbidib*. It can be given as doubles, triples or up to 10 notes in each trill. It also often repeats simple, short 'jip' notes in flight.

ALARM CALL In anxiety or alarm, typically gives a one-, two- or three-syllable-long whistle, with a very strong rise in pitch or rise-fall-rise, such as *dweee*, *dwee-ū* or *dwee-ū-ēe*.

SONG A series of rather melodic mini-trills, each separated by a short pause, and most based on the main calls. Once every few trills, it gives a long, slow wheeze, *dweeeeez*. In display flight, the male circles slowly, singing all the while.

COMPARE WITH Chaffinch (call); Brambling (song); Crossbill (call).

Chloris chloris

UK STATUS Declining but still abundant resident, with perhaps now around a million pairs, widely spread, augmented by some winter visitors from the continent.

HABITAT Gardens, especially with mature hedges and feeders, plus woodland and farmland. Some move onto vegetated beaches in winter.

Twite

The 'Linnet of the North' has a thick, fizzing call that gives it its name, plus a curious experimental song.

'TWITE' CALL The diagnostic call, *twī-eet*, *z'vee* or *zū-ree*, clearly rises in pitch, is relatively loud and has a thick, nasal quality.

TI-DIT CALL Be aware that the 'twite' calls are dotted among many very clipped *ti-dit* or *ji-ji-jit* calls, very similar to those of the Linnet.

SONG Verses can be long and unbroken but are often just half a second or so, each different and separated by short pauses. The 'notes' within each verse, whether in short or long song, are at a fairly relaxed pace and are full of ping and twang, twisting and warping. In many 'notes' you can hear two completely different sounds simultaneously; loafing flocks often sing in a group chatter.

COMPARE WITH Linnet (song and call); redpolls (call).

Linaria flavirostris

UK STATUS Resident and short-distance migrant, with around 10,000 pairs, most in north-west Scotland, especially around the coasts and islands. Small numbers survive in Ireland, Wales and the Peak District. Many move to the coast in the northern half of the country in winter.

HABITAT Breeds in coastal and in-bye meadows and on moorland, in winter some shifting to the seed-rich saltmarsh.

WHERE TO HEAR Try Mull, the Outer Hebrides, Orkney and Shetland in spring.

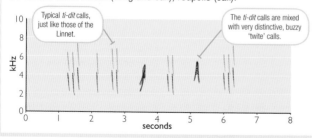

Typical *ti-dit* calls, just like those of the Linnet.

The *ti-dit* calls are mixed with very distinctive, buzzy 'twite' calls.

Linnet

Learning the *ti-dit* flight call is key, as it is woven into the exuberant twittering song of this gregarious farmland and heathland finch.

TI-DIT CALL The typical flight call, a fast, clipped *ti-dit* or *ti-di-dit*, the first note often slightly different in pitch, unlike redpolls, where each note in their *chi-chi-chit* is the same pitch.

SONG Pleasing, lively but very varied, some verses barely last 1.5 seconds, while 'long song' can be 15 seconds or so. The structure is much like Goldfinch song, a series of short trills linked with quick notes. These links include low buzzes and warped whistles that rise or fall sharply. The *ti-dit* call is the golden thread, often introducing and then dotted into verses. Resting flocks often sing and call in chorus.

COMPARE WITH Greenfinch (song and calls); Goldfinch (song and calls); redpolls (call); Twite (call).

Linaria cannabina

UK STATUS Widespread resident and summer visitor, with more than 400,000 pairs, especially common in coastal areas but scarce in the far north. Many migrate south to winter in western Europe and lowland areas.

HABITAT Breeds semi-colonially in short, dense scrub such as hedgerows and gorse, feeding together in weedy fields; scarce in gardens and urban areas.

Lesser Redpoll and Common Redpoll

These two charming and compact little finches – the Lesser (right) by far the commoner in the UK – are closely related with an identical repertoire.

CHI-CHI-CHI CALL The main call, frequently given in flight, is a confident *chi-chit* or *chi-chi-chi*, similar to the calls of the Linnet and the Twite but each note is 'thicker', and all at the same, rather deep pitch.

LONG CALL A strongly rising, long single whistle, *dsū'o'wee*, rather nasal and wavering, similar to the Siskin's call. It combines two different notes simultaneously, giving it an impure quality.

SONG Alternates the *chi-chi-chi* call with a long, straight rattle, *brrrrrrrr*, or more musical trill, *sirirɪrɪrɪ*, with short pauses in between each. It is typically given in undulating, wandering song flights. It sometimes adds the *dsū'o'wee* note into the sequence, and occasional longer songs have fewer pauses and varied trills.

COMPARE WITH Linnet (call); Siskin (call); Greenfinch (alarm call).

Acanthis cabaret
Acanthis flammea

UK STATUS The Lesser Redpoll is a resident and passage migrant, with more than 200,000 pairs, most in the north and west; some are fairly sedentary, others migrate to winter on the near continent. A handful of Common Redpolls breed in the far north of Scotland, and a few hundred from Scandinavia winter here.

HABITAT The Lesser Redpoll breeds in upland birch forests, young conifer plantations and lowland heathland. In winter, more widespread, often feeding in birches or streamside Alders, some visiting garden feeders.

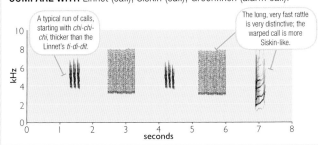

A typical run of calls, starting with *chi-chi-chi*, thicker than the Linnet's *ti-di-dit*.

The long, very fast rattle is very distinctive; the warped call is more Siskin-like.

Crossbill, Scottish Crossbill and Parrot Crossbill

Separating these three species is incredibly difficult, with bill size and shape important. The three species do differ in their calls but only subtly. Indeed, certain populations of Crossbill make unique calls and may be separate species!

FLIGHT CALL Given frequently in flight and when perched, in the Crossbill in Britain it is typically a sharp *chip* or *glip*, whereas the Parrot and the Scottish give a *choop* with strong, downwards inflection.

EXCITEMENT CALL Used to draw in other crossbills when excited or alarmed, it is deeper pitched than the flight call, typically *toop* in the Crossbill, a shorter *tup* in the Parrot and with a hint of ripple, *trup*, in the Scottish.

SONG Not often heard, it is a series of repeated clipped or seesawing notes, at about four notes per second, interspersed with calls.

COMPARE WITH Greenfinch (call).

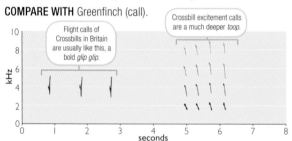

Flight calls of Crossbills in Britain are usually like this, a bold *glip glip*.

Crossbill excitement calls are a much deeper *toop*.

Loxia curvirostra
Loxia scotica
Loxia pytyopsittacus

UK STATUS The Crossbill is by far the commonest and most widespread of the three species but numbers vary, hugely determined by irruptions from north-east Europe and beyond. Scottish and Parrot Crossbills are sedentary residents in the Scottish Highlands, and the Parrot is also a rare visitor from north-east Europe. There are maybe 40,000 pairs of Crossbill, 7,000 of Scottish and 50 or so of Parrot.

HABITAT Large conifer forests. The Crossbill primarily feeds in spruce trees, the Scottish Crossbill and the Parrot Crossbill in Scots Pine.

Goldfinch

Track 36

You can hear the 'tickle it' calls and tinkling songs in many gardens. See page 77 for full description.

'TICKLE IT' CALL Frequently heard, the incredibly short, fast *tik-a-lit* notes bounce about all over the place, often dotted with quiet, high *sik* notes.

OTHER CALLS Gives longer, pitch-bending calls, such as *z^{re}e* or *d^{see}oo*.

THREAT CALL In confrontations, gives livid, loud *chrrt chrrt* calls.

SONG Can seem a random, tinkling twittering, with verses short or extended to 10 seconds or more. However, listen how *tik-a-lit* sounds link together fast, bright trills and connecting longer notes. It all flows without pause, unlike the Greenfinch song. Groups of males sing communally, especially before roosting and in autumn.

COMPARE WITH Linnet (song and call); Siskin (call); Swallow (call).

Carduelis carduelis

UK STATUS Abundant and widespread resident and summer visitor, with about 1.2 million pairs. Many winter in south-west Europe but some just wander locally in winter.

HABITAT Needs abundant fresh seeds of weeds in spring and summer, so is found along field margins, rough ground and pasture, often in and around villages but now also in urban centres. In gardens in winter.

Siskin

As acrobatic as a tit, this dainty, tree-loving finch is a frequent voice in many conifer forests.

WARPED CALL This frequent contact call is given freely in flight. It is a two-syllable, off-key $d\bar{u}'e^e$ or $d_{e_e}'\bar{u}$, warping up or down as if it has a fault in the middle. The redpolls' $ds\bar{u}'o\,'w_{e^e}$ is similar but longer.

CHITTER CALL A pitter-patter of very fast *cht cht cht* notes that can turn into louder, angry threat calls.

SONG Given from treetops or in slow, circling display flight, each verse is a fast shuttling run of sticky little notes for many seconds, without the trills of the Goldfinch or Linnet and more like Swallow song. Some include short imitations of other bird calls. Towards the end, it inserts a long, nasal *dzweeeeee*. Groups of males often sing together, creating quite a din!

COMPARE WITH Goldfinch (song and call); redpolls (call).

Spinus spinus

UK STATUS Resident, with more than 400,000 pairs, many moving south or to lower altitudes in winter. It is also a passage migrant and winter visitor from northern Europe in varying numbers.

HABITAT Largely found in conifer forests in the breeding season, especially spruce, and feeds in streamside birches and Alders in winter. Tends to visit gardens when natural seed supplies are exhausted.

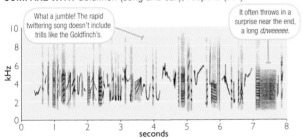

What a jumble! The rapid twittering song doesn't include trills like the Goldfinch's.

It often throws in a surprise near the end, a long *dzweeeee*.

Corn Bunting

Track 83

This chunky bunting of arable farmland has a song often likened to shaking a big bunch of keys.

SONG See page 116 for full description. Males often sing well into summer. Delivered from a low vantage point, each verse lasts about 1.5–2 seconds, starting with an accelerating rising series of five to eight simple clipped *pit* or *kik* notes, then breaking into a swirling metallic jangle. The 'shaking keys' element lasts for about a second, and then it tends to quickly peter out.

CALL The most typical call is a very clipped and rather high-pitched *prit! prit!*, given by anxious birds when perched or on take-off and in flight, sometimes run into *prit-it-it*, and often given in chorus by flocks. Listen, too, for longer, ratchety ^{skir}l or $s^{keer}t$ notes.

COMPARE WITH Yellowhammer (call).

Emberiza calandra

UK STATUS Sedentary resident, with around 10,000 pairs, most in England on downland from Dorset up through Wiltshire and the Chilterns to Cambridgeshire, and with some on the east Scottish coast and Outer Hebrides. It is absent from Wales and Ireland.

HABITAT Breeds mainly on chalky soils, especially around spring-sown crops and weedy areas; in winter it gathers in flocks to forage in stubble fields.

Yellowhammer

Much declined, there is nevertheless plenty of opportunities to hear the brightly coloured male's famous song in farmland hedgerows.

SONG See page 115 for full description. Each verse is a leisurely jogging rattle, finishing with a longer buzz, widely known as 'a little bit of bread and no cheese'. Sometimes there are actually two end notes – '… cheese, please' – whereas some verses omit the cheese.

THICK CALL The main contact call is a thick, smudgy *chidd*, so vague that others have transcribed it as *stüff*, *trlp* or *tswik*. It can also give a shorter and sharper *chit*.

OTHER CALLS In alarm, gives a little ripple, *sirut* or *chillup*, that drops in pitch, rather like that of the Long-tailed Tit. It also has a fine, thin, falling *seeee* call, more sibilant than the Reed Bunting's.

COMPARE WITH: Cirl Bunting (song and call), Lesser Whitethroat (song); Long-tailed Tit (call); Reed Bunting (call).

Emberiza citrinella

UK STATUS Sedentary and widespread resident, with around 700,000 pairs, but absent from the far north and now scarce in Northern Ireland.

HABITAT Predominantly a bird of farmland, especially with thick hedgerows and wildlife margins. It is also found on lowland heaths. In winter, flocks gather to feed in stubble fields.

Cirl Bunting

This speciality of coastal south-west England is closely related to the Yellowhammer, and has a simple straight rattle song.

SONG The male sings from near the top of a bush or hedge, each verse a fast, even rattle lasting about 1–1.5 seconds, *jnk-jnk-jnk-jnk-jnk-jnk*…. Each of the 12–30 'notes' is actually an ultra-fast couplet, giving the sound its slightly metallic or 'chinking' feel. Different males sing at different speeds, the slowest at about six couplets per second, similar to the Yellowhammer's rattle, the fastest at 20 per second, almost like a Grasshopper Warbler. Lesser Whitethroat song is similar but Cirl Bunting verses tend to be longer, with no intro, and with a more insect-like, mechanical feel.

CALLS The most typical call is a very high-pitched, short and feeble *sip*, dropping in pitch. In anxiety, it may extend into more of a *tzee*.

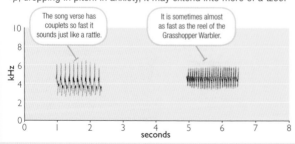

The song verse has couplets so fast it sounds just like a rattle.

It is sometimes almost as fast as the reel of the Grasshopper Warbler.

Emberiza cirlus

UK STATUS Sedentary resident, with the population now exceeding 1,000 pairs, but rarely seen outside south Devon or the recent reintroduction area in south Cornwall.

HABITAT Sheltered hedgerows, scrubby areas and field margins, often within sight of the sea, feeding in weed-rich and stubble fields.

WHERE TO HEAR Labrador Bay and Prawle Point (Devon).

Reed Bunting

A reedbed specialist, the dapper male, with his jet-black head and white moustache, sings two versions of his 'counting song' depending on his marital status! See page 125 for full description.

SLOW SONG Given by mated males, each verse is three to seven plodding notes, as if the bird is tentatively learning to count. So one verse might run to the rhythm of 'one one two four', the next 'one four two-two three'.

FAST SONG Unmated males use the same range of notes but maybe 8–12 per verse at a much faster six notes per second, and so sound more accomplished.

MAIN CALL Imagine saying the name 'Sue' as thinly as you can through tightly pursed lips, with a strong drop in pitch – $^{t}si\bar{u}$.

ALARM CALL A straight-pitched, rather thick $tch\bar{u}$, often given in flight.

COMPARE WITH Lapland Bunting (call); Yellowhammer (call).

Emberiza schoeniclus

UK STATUS Widespread resident, with about 250,000 territories, mainly sedentary but dispersing locally in winter.

HABITAT Breeds in well-vegetated marshes and fens, especially reedbeds, plus young conifer plantations and fields of oil-seed rape. In winter, it seeks out seed-rich habitats such as stubble fields but still roosts in wetlands. It visits a few bird tables in winter.

Lapland Bunting

Rarely seen, this bunting behaves like a mouse, creeping among short vegetation. Many sightings are of flyover migrant birds as they pass along the coast, identified by their calls.

FLIGHT CALLS The varied calls include a clipped *tirit* or *tiririt* like that of the Snow Bunting, a downslurred, thick $^{t}ch\bar{u}$ and a thinner *chup*.

COMPARE WITH Snow Bunting (call).

Calcarius lapponicus

UK STATUS Scarce winter visitor to the east coast from northern Europe.

HABITAT Coastal fields in winter, especially in weedy stubbles.

Snow Bunting

This chunky bunting, flashing white as it flies, enlivens a visit to a bleak winter beach or our very highest mountain tops in summer.

CALL There are three main calls. Most typical is an attractive ripple, *tiririp*, meandering in pitch and rather like that of the Turnstone. Also has a simple $^{t}y\bar{u}$ whistle, and a deep buzz, *spurj*, in alarm.

SONG The simple 1.5–2.5-second verses are relaxed little melodies, rich and mellow, at about six notes per second, which wander up and down with little repetition.

COMPARE WITH Lapland Bunting (call); Shore Lark (call).

Plectrophenax nivalis

UK STATUS Around 80 pairs breed in the Scottish mountains; otherwise, about 10,000–15,000 visit Scotland and the English east coast in winter.

HABITAT Breeds on high mountain tops. In winter, northern moors, coastal stubbles and beaches.

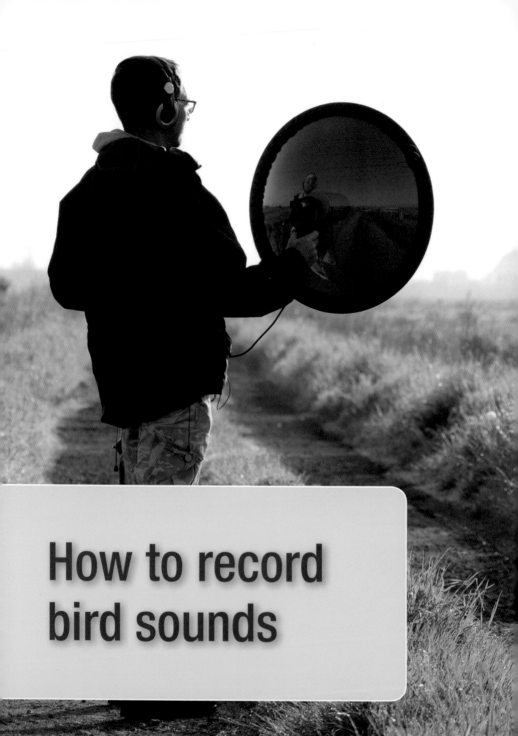

How to record
bird sounds

How to make your own recordings

This book may have made you think of recording bird sounds for yourself, but how do you go about it?

At its simplest, you only need a microphone and a device to record on, and most modern mobile phones can be both of those things. Using your phone is an easy way of grabbing a reminder of what you've heard in the way you might take a 'record shot' with a camera. Your recording won't be broadcast quality, but it will give you something to play back and compare with the recordings and text in this book.

However, if you want to make higher quality recordings, you will need:

- a microphone ('mike'), with associated grip, cables and windshield;
- a sound recorder with memory cards and batteries;
- earphones.

Here is a guide to these in more detail.

Microphones

Some bird sound recordings capture the overall ambience of a location, for which an omnidirectional microphone is used to pick up sound from all directions. The purpose of most recordings, however, is for the recordist to create a record of individual birds. The human brain is very good at focusing on one sound and editing out all others, but to achieve that with a mike, recordists need one of two main types of kit: shotgun mikes and parabolic reflectors.

Shotgun mikes

These are thin, light metal tubes, typically about 30cm long and 2.5cm in diameter. You will often see them covered with fluffy windshields (which I call 'shaggy dogs'; below). Shotgun mikes do not amplify sounds but exclude those on either side. They are relatively light and easy to transport.

Parabolic reflectors

These use small, often omnidirectional mikes, but they sit in the middle of what looks like a large plastic satellite dish, typically about 60cm wide (left). The dish focusses the sound coming towards it onto the microphone, amplifying it. It doesn't completely eradicate noises from other directions, and of course it also magnifies other sounds in the direct line of your subject.

To help diminish the sound of wind whirling around the reflector, you can also buy the dish its own fluffy

cover. The dishes are light but somewhat awkward when pushing through vegetation or clambering over stiles, and they get very buffeted by the wind. Oh, and people are fascinated by them! While using one, I have been asked – in all seriousness – if I am trying to tune in to a TV station.

Parabolic reflectors are also more expensive, and they amplify higher-pitched sounds more than those at a low pitch. However, they create excellent quality recordings and overall are the preferred equipment for most recordists. They are what I used to record the sounds on the audio that accompanies this book.

Recording devices

A broad range of recording devices is available, from simple and reasonably priced equipment (shown above) to a more professional level of kit. Many modern devices that are perfectly adequate for the serious amateur are also small enough to fit in a pocket.

Most models will have the basics, which will include:

- a mike input ('mic'), which will either be a plug-in power (PiP) jack or a three-pin XLR socket;
- a headphone ('ear') socket;
- a slot for digital storage SD ('secure digital') cards.

You should also make sure that your device:

- has a display screen that lights up for use in low light conditions;
- has the ability to set the recording level manually;
- and can record in WAV or PCM format. Ideally, it should be able to record in CD format (which is uncompressed PCM audio at a 44.1kHz sample rate).

You need to be able to hear your recording as it is being made, so you should connect your device to headphones. If portability is important to you, it is possible to use ear-bud types quite successfully, but on-ear or over-ear headphones are preferable to dampen extraneous noise, although they are much bulkier.

The recorder is likely to use AA-type batteries. Just be aware that recorders are quite energy hungry, so always carry spares.

How much will it cost?

The answer to that depends on your goals, your budget and how long it is since publication that you read this. To give you an idea, at 2019 prices, you'd expect to pay maybe £400–£1,000 for a good-quality, entry-level microphone (a parabolic reflector kit will be at the top end of that), £100–£500 for a decent recorder, plus cables, digital cards and batteries.

It means that a budget of £600–£1,500 will allow you to make decent quality recordings. If, however, you would like to achieve excellent rather than good-quality recordings, the sky is the limit in terms of cost.

Field recording technique

Ok, you've got a mike, recorder and earphones, and you're ready to make your first recordings. Here are some tips to help you get the best results.

1. **Choose a quiet time and place to record**.
   Sadly, the first time you record you will find out what a noise-polluted world we live in, and it affects almost every recording session. Cars, lawnmowers, extractor fans, dogs barking, transistor radios, pneumatic drills, trains, vehicle-reversing beeps...

   It isn't just a problem in our towns and cities. Out in the countryside, there are so many places you'd think would be free of noise but you turn up only to encounter a distant generator, or the hum of a power station, or the pulsing feedback from a radio transmitter aerial. Recording at dawn in my home village in rural Worcestershire is incredibly difficult due to the continuous hum of the sewage farm pumps and the growl of the M5 motorway 2 miles away as the crow flies.

   My three worst sound frustrations? Motorbikes, farm machinery (not because they are any noisier than other machinery but because they are often in action at dawn) and aircraft.

   Just when you think you have got away from all the intrusive human noise, the wind gets up and rustles the leaves, or you can hear the sea pound a distant beach, or it starts to rain. Oh, and as for sheep...

   So sound recording is often best done in remote (sheep-free) locations and very early in the morning, which has the added benefit that it is usually when the air is at its most still. However, remember that the dawn chorus can be so jammed with birdsong that it can be difficult to record one bird in isolation.

2. **Check your equipment**. Do you have spare batteries and enough space on your memory cards? Do you have warm clothing and supplies, for recording bird sounds takes time and much standing around? Are you wearing rustle-free clothing? Right, you're ready to head into the wilds.

3. **Now, find a bird that is making a noise**. It could be that you know of a regular song post or it could be just a chance encounter. So earphones on, switch the recorder on and press the record button, which will typically take you into standby mode, allowing you to hear what the mike is picking up. Point the microphone towards your subject – this may sound obvious, but it isn't always easy. Getting a visual fix on the bird can really help; otherwise, it is a case of slowly waving the mike until you hit the 'sweet spot'.

4. **Adjust the recording level** ('gain') so that it is as loud as possible but without distorting the recording. Most recorders will have a screen to allow you to monitor this.

5. **Now press the record button, and stand *very* still**. It is easy to create what is called 'operational noise' while recording, especially creaking sounds as you move the mike, but heavy breathing or your belly gurgling will also be registered!

6. **Now, with an initial recording in the bag, you might like to work your way closer**, using all your field skills, adjusting the recording level as you go. Move very slowly, and at an angle to the bird rather than directly towards it, pausing often. The closer you are, the more you'll be able to turn down the gain and the quieter any background noises will become. But remember, the bird's welfare comes first.

7. **Before you finish recording**, say quietly into the mike what bird you were recording (if you know) and where you are. You will find this invaluable when going back through your audio files later. Press the stop button and check the recording has stopped (I have failed on this last point many times and found I have then captured the sounds of me walking!).

The ethics of playing sounds in nature

One technique that some people use when recording a bird is to play a short recording of its song out loud. This can dupe a male into thinking there is an intruder in his territory, prompting him to come closer and to sing furiously. This technique is called 'playback'.

However, there is an ethical angle here, and in some cases a legal element, too. As the Birdwatchers' Code of Conduct says:

'Repeatedly playing a recording of bird song or calls to encourage a bird to respond can divert a territorial bird from other important duties, such as feeding its young. Never use playback to attract a species during its breeding season.'

'In England, Scotland and Wales, it is a criminal offence to disturb, intentionally or recklessly, at or near the nest, a species listed on Schedule 1 of the Wildlife & Countryside Act 1981. Disturbance could include playback of songs and calls. The courts can impose fines of up to £5,000 and/or a prison sentence of up to six months for each offence.'

So it is possible within the Code to use playback outside of the breeding season, but I do not use it to make any of my recordings, and I enjoy the challenge of doing it that way.

Post-production

Once you have some recordings 'in the bag', it is good practice to start by logging your recordings. In a quiet room, listen to each track, noting down which species occur at which times.

You may then want to edit your audio files, deleting poor recordings or cutting out unwanted sections. For this, you will need a computer and some software. There are many professional editing packages, but freeware is readily available and entirely sufficient for simple cutting and splicing of sounds.

You might also like to try some simple 'cleaning up' of your recordings, which for bird sounds usually involves removing deep-pitched background noise using what is called a high-level filter. For example, setting the filter at 100Hz will remove any sounds at that pitch or below. However, there are many background noises that you just won't be able to do anything about, which confirms the importance of getting as clean a recording as possible in the first place.

For much more information, guidance and advice on all the aspects above, go to the website of the Wildlife Sound Recording Society www.wildlife-sound.org.

xeno-canto

The recordings that accompany this book contain more than 100 different types of common bird songs and calls from 65 species. But how can you hear the other, more unusual sounds? Help is at hand because you can listen to almost any bird sound you want, from anywhere in the world, courtesy of the excellent and free-to-access website www.xeno-canto.org, set up by Bob Planqué and Willem-Pier Vellinga in 2005. Their aims are to:

- popularise bird sound recording worldwide;
- improve accessibility to bird sounds;
- increase knowledge of bird sounds.

They have done just that: by 2019, the database was rapidly approaching 500,000 recordings from about 5,000 people and includes all the species described in Part 3 of this book.

Glossary

Advertising call Typically used to describe the particular calls that non-passerines use to attract mates. These calls are similar to the songs of passerines but advertising calls are usually simpler in structure.

Antiphonal duetting In which one bird in a pair sings, prompting its mate to sing immediately in response, creating alternating bursts of song.

Call A vocal sound other than a song often used to signal intent or action. For different types of call used in specific situations, see pages 12–13.

Couplet A two-note phrase repeated within a song verse.

Crescendo The gradual increase in volume within a note or a series of notes.

Crystallised song The consistent and small repertoire of verses that mature songbirds settle upon.

Dawn chorus The combined songs of birds at dawn, usually at its peak in spring.

Decay The time it takes for a sound to fade to silence.

Dialect A song or call particular to a geographical population of birds.

Display flight A ritualised flight pattern a bird uses to attract a mate or defend its territory.

Disyllabic A note with two clear syllables.

Doppler effect Whereby a sound from a source coming towards or moving away from you causes a sudden and noticeable change in pitch.

Dread When a chorus of bird calls falls instantly silent.

Drum/Drumming The rapid volley of beak strikes against a hard surface, as made by woodpeckers. The sound produced by the vibrating outer tail feathers in the display flight of the Snipe.

Duet The sound of a pair of birds singing simultaneously.

Duration The length of a sound (usually given in seconds).

Fundamental pitch The dominant pitch in a sound, often but not always the lowest pitch.

Harmonics Evenly spaced layers of pitch in a sound that give it a rich, nasal quality.

Imitation When a bird copies other sounds it hears, usually those made by another bird.

Inharmonics Overtones of a sound that are not even multiples of the fundamental pitch.

Note Individual sounds, the building blocks of calls and songs.

Overtones All the pitches within a sound other than the fundamental pitch.

Passerines A group (Order) of birds often known as the 'perching birds'. In a British and European context, the term is used synonymously with 'songbirds'.

Phonetic transcription The approximate rendering of a bird sound as 'words'.

Phrase A distinct section within a longer verse.

Plastic song The variable phrasing used by a young bird as it develops its crystallised song.

Pure notes A sound without overtones, which typically creates a clear, whistling sound.

Rattle A rapid series of the same percussive note, in which one can hear each note, but they are too fast to be counted.

Repeat series A series of the same note, slow enough to count.

Repertoire The overall sum of all the different calls and verses that one individual or species makes.

Roding The dusk display flight of the Woodcock.

Seesawing In which two notes of different pitches alternate repeatedly.

Song A special type of vocalisation, mainly given by males to attract a mate and defend a territory, and primarily applied to passerines (songbirds).

Songbirds Technically the Oscine passerines; in a British and European context, used synonymously with 'passerines'.

Song flight A ritualised flight pattern a songbird uses to accompany its song.

Song type A group of verse types that are similar in overall feel; most birds have one song type, but a few have two very different song types.

Sonogram A visual representation of the form of a sound over time, plotted against the pitch.

Sparrow chapel The name for a vocal group of sparrows when gathered habitually in a favoured place.

Staccato Notes that are short and clipped, with a pause between each note.

Stereotyped Vocalisations which show little variation across a species.

Subsong The subdued practice song of a bird, usually quiet, rambling and without a clear structure.

Syllable A note with one vowel sound.

Syrinx The organ of vocal sound in birds, located at the base of the windpipe.

Trill A rapid series of the same musical note, in which one can hear each note, but they are too fast to be counted.

Triplet A three-note phrase repeated within a song verse.

Verse One single burst of song, with a definite start and end.

Verse types When a bird has two or more consistently structured song verses that it switches between.

Vibrato A sound where the pitch wavers rapidly up and down.

Whipped/whizzing The quality of a sound that leaps rapidly in pitch, up, down or both.

Yodel A sound that breaks sharply from a low to high pitch, or vice versa.

References

Balmer, D., Gillings, S., Caffrey, B., Swann, B., Downie, I., & Fuller, R. 2013. *Bird Atlas 2007–11*. BTO Books.

Constantine, Mark & The Sound Approach. 2006. *The Sound Approach to Birding*. The Sound Approach.

Cramp, S. (Chief Ed.) 1994. *Handbook of the Birds of Europe, the Middle East and North Africa. Vols I–X*. Oxford University Press.

Ehrlich, P., Dobkin, D.S., Wheye, D. & Pimm, S.L. 1994. *The Birdwatcher's Handbook*. Oxford University Press.

Farrow, D. 2008. *A Field Guide to the Bird Songs & Calls of Britain and Northern Europe*. Carlton Books.

Musgrove, A.J., Austin, G.E., Hearn, R.D., Holt, C.A., Stroud, D.A. & Wotton, S.R. 2011. Overwinter population estimates of British waterbirds. *British Birds*, 104: 364–397.

Sample, G. 1998. *Bird Call Identification*. HarperCollins.

Svensson, L., Mullarney, K. & Zetterstrom, D. 2010. *Collins Bird Guide*. Collins.

Acknowledgements

This book and recording follow humbly in the footsteps of those who pioneered the art of recording bird sounds which is still less than 100 years old.

In Europe, the work of Jean C. Roche in the 1950s and 1960s was groundbreaking, and it is astonishing how he managed to record so many bird species when there wasn't information widely available about where to find them all.

Later influential works include Dominic Couzens' and John Wyatt's *Teach Yourself Birdsounds* series of audio cassettes, and books and CDs by Geoff Sample and Dave Farrow. A further leap forward came from the team who call themselves The Sound Approach led by Mark Constantine, Magnus Robb and Arnoud van den Berg. Their detailed, sophisticated look at the subject has resulted in many new insights. To all of these people, I owe a huge debt.

On a more personal level, my thanks are due to Duncan MacDonald at WildSounds for introducing me to the world of bird sound books, CDs and tapes, and who has done so much over 30 years to champion the subject. Thanks too to Julie Bailey at Bloomsbury for seeing the potential in this project, to Jenny Campbell for steering me through the complex editing process, to Julie Dando for her design flair, and to Julian Baker for his work formatting the sonograms. My thanks to the RSPB for allowing me to spend a glorious four-week sabbatical in 2016, recording birdsong on RSPB nature reserves. My thanks, too, to Dr Richard Black for supporting my first tentative steps with microphone in hand, to Liam McMillan at Small Pond Studios for his recording studio expertise and attention to detail, and to Peter Boesman, Malcolm Shaw and David Darrell-Lambert for the use of their sound recordings. My thanks, too, to Mum and Dad for filling my childhood with birdsong, and to Peter Francis for his support and patience.

Sound Credits

All sound recordings by Adrian Thomas except for the following by Peter Boesman: Goldcrest call (29), Starling song (40, 45), White Wagtail chizzik call (43), Willow Warbler call (57), Nuthatch call (62), Green Woodpecker song (63), Marsh Tit call (66, 73), Long-tailed Tit call (68), Spotted Flycatcher call (72), Meadow Pipit call (88); and Malcolm Shaw: Bullfinch call (70, 73). All track numbers are shown in parentheses.

Photo Credits

Sonograms devised and drawn by Adrian Thomas, with electronic files created by Julian Baker.

Bloomsbury Publishing would like to thank the following for providing photographs and for permission to reproduce copyright material within this book. While every effort has been made to trace and acknowledge all copyright holders, we would like to apologise for any errors or omissions and invite readers to inform us so that we can make corrections at future editions.

Key: t = top; tl = top left; tr = top right; tc = top centre; c = centre; b = bottom; bl = bottom left; br = bottom right. AL = Alamy; AT = Adrian Thomas; DT = David Tipling/birdphoto.co.uk; GI = Getty Images; IS = iStock; NPL = Nature Picture Library; RS = RSPB Images; SS = Shutterstock.

Front cover c Chris Gomersall/RS, tr Aleksandrs Bondars/SS; **back cover** tl Steve Round/RS, tc Erni/SS, tr Roger Wilmshurst/RS; **CD label** c Chris Gomersall/RS; **2** Matthew Cattell/GI; **3** r bearacreative/SS, c Ttphoto/SS, l Sandra Standbridge/SS; **6–7** Wilfried Martin/GI; **8** Paul Hobson/NPL; **9** Christine Hoi/SS; **10** miha de/SS; **12** DT; **13** Michel Rauch/GI; **14** Sven Zacek/GI; **15** Wildlife World/SS; **16** AT; **20** Menno Schaefer/SS; **21** CezaryKorkosz/SS; **24** Vlasto Opatovsky/SS; **25** Nature Picture Library/GI; **30** Frank Fichtmueller/SS; **32** Andy Sands/NPL; **33** Kevin Sawford/NPL; **35** Julie Dando, Fluke Art; **36** tl Taiga/SS, tr Skreidzeleu/SS, wat/IS, Simon Bratt/SS; **37** t Marina Zezelina/SS, b Mike_Pellinni/IS, night Marina Zezelina/SS, dawn MudMee/SS, sunrise bunsview/SS, mid-morning kwasny221/IS, midday Mike_Pellinni/SS, mid-afternoon Finding Horizons/SS, sunset r.classen/SS, dusk Zacarias Pereira da Mata/SS; **38** Mike Kemp/Contributor/GI; **39** Altitude Drone/IS; **40–41** Adam Burton/AL; **42–43** skyfilming.com/GI; **44** Andy Hay/SS; **45** Peter Cox/SS; **46** David Tipling/RS; **47** Colin Wilkinson/RS; **48–49** David Wootton/RS; **50** Loop Images/Craig Joiner/GI; **51** Andy Hay/RS; **52–53** photographie de paysages-/GI; **54** Nigel Blake/RS; **55** AT; **56–57** Wilfried Martin/GI; **58–59** Malcolm Hunt/RS; **60** t Ian_Redding/IS, b Mirko Graul/IS; **61** Andrew_Howe/IS; **62** t Menno Schaefer/SS, b lues01/IS; **63** tony mills/SS; **64** t Chris Gomersall/RS, b Richard Bowler/RS; **65** t Erni/SS, b Leopardinatree/IS; **66** t Ralph DeseniÃ/GI, b Andrew_Howe/IS; **67** t Wolfgang Kruck/SS, b schnuddel/IS; **68** t Michal Pesata/SS, b Piotr Krzeslak/SS; **69** Kaleel Zibe/RS; **70** t David Tipling/NPL, b Voodison328/SS; **71** t Lifeafterwork1/IS, b Andrew_Howe/IS; **72** t Fireglo/SS, b mzphoto11/IS; **73** t ViktorCap/IS, b gregg williams/SS; **74** t Ian Bray/EyeEm/GI, b Joe Ravi/SS; **75** t Bachkova Natalia/SS, b TT/IS; **76** t Ernie Janes/NPL, b Ysbrand Cosijn/IS; **77** t Rafal Szozda/SS, b Genevieve Leaper/RS; **78** Arto Hakola/SS; **79** t Dave Montreuil/SS, **b** Mark robert paton/SS; **80** t lightix/IS, b Franke de Jong/SS; **81** t Anton Luhr/GI, b Erni/SS; **82** t Andrew_Howe/IS, b gallinago_media/IS; **83** t DT, b DT; **84** Ian D Nicol/SS; **85** t HERGON/SS, b Hubert Schwarz/SS; **86** t Jevtic/IS, b Kevin Sawford/RS; **87** Oleg Troino/SS; **88** t Chris Gomersall/RS, b shaftinaction/SS; **89** Jan Sevcik/RS; **90** t Daniel Zuppinger/SS, b Andrew_Howe/SS; **91** t Mateusz Sciborski/SS, b Michael Schroeder/SS; **92** t Gerdzhikov/IS, b Andyworks/IS; **93** t DT, b MikeLane45/IS; **94** t David J Slater/RS, b MikeLane45/IS; **95** bearacreative/SS; **96** t Piotr Krzeslak/SS, b Piotr Krzeslak/AL; **97 t** Giedriius/SS; **97 b** Mark Medcalf/SS; **98** t StockPhotoAstur/IS, b Jerome Murray – CC/AL; **99** t Piotr Krzeslak/SS, b Guy Rogers/RS; **100** t cimbat2/SS, b Jurgen & Christine Sohns/GI; **101** t Sandra Standbridge/SS, b Erni/SS; **102** t Ondrej Prosicky/SS, b charliebishop/IS; **103** t suerob/IS, b Ray Kennedy/RS; **104** t Erni/SS, b Sue Robinson/SS; **105** t Mark Hamblin/RS, b hardeko/IS; **106** t Sylvia Adams/SS, b Oliver Smart/SS; **107** ajt/IS; **108** t Steve Round/RS, b Andrew_Howe/IS; **109** t Erni/SS, b Andyworks/IS; **110** t Chris Gomersall/RS, b Erni/SS; **111** Wildlife World/SS; **112** t AbiWarner/IS, b Friedhelm Adam/GI; **113** t Andrew_Howe/IS, b VictorTyakht/IS; **114** t Mike Lane/RS, b Roger Tidman/RS; **115** t garmoncheg/SS, b M Rose/SS; **116** t xpixel/SS, b Salparadis/SS; **117** t chris2766/IS, b Menno Schaefer/SS; **118** t Alexander Erdbeer/SS, b Volodymyr Kucherenko/IS; **119** t Alexey Lesik/SS, b Drakuliren/SS; **120** t kaeja2525/SS, b Piotr Krzeslak/IS; **121** t MikeLane45/IS, b Ger Bosma/GI; **122** t Tobyphotos/SS, gui00878/IS; **123** t Ruth Black/SS, b Paolo-manzi/SS; **124** DT; **125** t Erni/SS, b SoopySue/IS; **126** t DT, b Juniors Bildarchiv GmbH/AL; **127** t Mark L Stanley/GI, b Chamara SKG/GI; **128** t MikeLane45/IS, b DT; **129** WMarissen/IS; **130–131** Keith Pritchard/SS; **132** t DT, b Brian Lasenby/SS; **133** t Nick Pecker/SS, b Red Squirrel/SS; **134** tl aseppa/IS, tr AGAMI Photo Agency/AL, b DT; **135** t DT, b DT; **136** t DT, b Erni/SS; **137** t WOLF AVNI/SS, b DT; **138** t Ronald Wittek/SS, c DT, b DT; **139** t DT, b DT; **140** t M Rose/SS, b DT; **141** VV Shots/SS; **142** t DT; **143** DT; **144** t DT, b DT; **145** DT; **146** t DT, b DT; **147** t DT, b DT; **148** t DT, b Wildlife World/SS; **150** t DT, b DT; **151** t DT, b Richard Steel/BIA/Minden Pictures; **152** t DT, b DT; **153** t Simonas Minkevicius, b DT; **154** Arto Hakol/SS; **155** t DT, b DT; **156** DT; **157** t DT, b DT; **158** t David Tipling/GI, b A.S.Floro/SS; **159** t Stubblefield Photography/SS, b Dennis Jacobsen/SS; **160** t Jesus Giraldo Gutierrez/SS, b Peter Louwers/SS; **161** Menno Schaefer/SS; **162** DT; **163** t Piotr Krzeslak/SS, b DT; **164** t MikeLane45/IS, b DT; **165** t Wildlife World/SS, b Miroslav Hlavko/SS; **166** t DT, b DT; **167** t DT, b Bildagentur Zoonar GmbH/SS; **168** t CyberKat/SS, b DT; **169** DT; **170** t Erni/SS, b Mirko Graul/SS; **171** t DT, b DT; **172** t DT, b Vitaly Ilyasov/SS; **173** Erni/SS; **174** t MikeLane45/IS, b DT; **175** aaprophoto/IS; **176** t David J Slater/RS, b DT; **177** DT; **178** t FLPA/Paul Sawer/GI, b Mark Medcalf/SS; **179** Andrew M. Allport/SS; **180** t DT, b Keith Pritchard/SS; **181** t Voodison328/SS, b DT; **182** t DT, b Jukka Palm/SS; **183** t RazvanZinica/SS, b skapuka/SS; **184** t Keith Pritchard/SS, b Chris Moody/SS; **185** t Voodison328/SS, b Erni/SS; **186** t Wolfgang Bittermann/SS, b Chris Schenk/Buiten-beeld/Minden Pictures/GI; **187** t Jerome Whittingham/SS, b Marcel van Kammen/NiS/Minden Pictures/GI; **188** t DT, b DT; **189** t raulbaenacasado/SS, b hstiver/IS; **190** t DT, b DT; **191** t Gallinago_media/SS, c DT, b Menno Schaefer/SS; **193** t DT, b fernando sanchez/SS; **194** t DT, b Adam Fichna/SS; **195** t DT, b DT; **196** t Sokolov Alexey/SS, hfuchs/SS; **197** t taviphoto/SS, c Daniele Occhiato/Buiten-beeld/Minden Pictures/GI, b Ivan Godal/SS; **198** t DT, b Tone Trebar/SS; **199** t aaltair/SS, Paolo-manzi/SS; **200** Red ivory/SS; **201** Martin Mecnarowski/SS; **202** t Jesus Giraldo Gutierrez/SS, b Erni/SS; **203** t Piotr Krzeslak/SS, b Erni/SS; **204** t John Navajo/SS, b DT **205** t Rudmer Zwerver/SS, bl Florian Teodor/SS, br Rudmer Zwerver/SS; **206** t DT, b Davydele/SS; **207** t J.M.Abarca/SS, b Bachkova Natalia/SS; **208** Szczepan Klejbuk/SS; **209** t Bachkova Natalia/SS, b Richard Constantinoff/SS; **210** Wildlife World/SS; **211** DT; **212** t charlathan/IS, b DT; **213** t DT, b Georgios Alexandris/SS; **217** t DT, b DT; **218** t DT, b Vitaly Ilyasov/SS; **219** t rock ptarmigan/SS, CezaryKorkosz/SS; **220** t DT, b Erni/SS; **221** t DT, b Ondrej Chvatal/SS; **222** t Erni/SS, b DT; **223** t Giedriius/SS, b LABETAA Andre/SS; **224** t DT, b DT; **225** t DT, b Martin Fowler/SS; **226** t ArCaLu/SS, b Richard Steel/BIA/Minden Pictures/GI; **227** t Wildlife World/SS, b Wolfgang Kruck/GI; **228** t DT, b DT; **229** t DT, b Soru Epotok/SS; **230** t DT, b Sandra Stanbridge/SS; **231** t DT, b DT; **232** t DT, b DT; **233** t Erni/SS, b Utopia_88/SS; **234** t DT, b Vishnevskiy Vasily/SS; **235** t allanw/SS, b Jesus Giraldo Gutierrez/SS; **236** t Ondrej Prosicky/SS, b Sandra Stanbridge/SS; **237** t DT, b Mark Caunt/SS; **238** t clarst5/SS, b DT; **239** t Ivan Godal/SS, b serkan mutan/SS; **240** t DT, b Erni/SS; **241** DT; **242** AT; **243** AT; **244** AT.

Track List

Below is a list of all the tracks featured on the narrated recording. If you are unable to play CDs or have any issues saving the audio from the CD to your computer, you can also download the audio from the CD here: www.bloomsbury.com/rspb-guide-to-birdsong. In particular, the track list sets out which species appear in the 'Over to you' medleys, so that you can check if you have correctly identified each bird sound.

1   Introduction
2   One Minute of Bird Sound
3   Terminology
4   Mystery Bird
5   Duration
6   Pace
7   Volume
8   Pitch
9   Pattern
10 Timbre
11 Phonetic Transcription
12 Overall Effect
13 Mnemonics
14 Mystery Bird
15 Blackbird Song
16 Robin Song
17 Medley: Robin song, Blackbird song (at 30 seconds); the two then alternate for several verses.
18 Wren Song
19 Dunnock Song
20 Medley: Dunnock song, Wren song, Dunnock song again (for a couple of verses), Wren song, Dunnock song.
21 Blackbird and Robin Calls
22 Wren and Dunnock Calls
23 Great Tit Song
24 Blue Tit Song
25 Coal Tit Song
26 Goldcrest
27 Medley: Blue Tit song, Goldcrest song, Coal Tit song (at 28 seconds), Great Tit song (at 39 seconds), Blue Tit song (again).

28 Great Tit and Blue Tit Calls
29 Coal Tit and Goldcrest Calls
30 Woodpigeon Song
31 Collared Dove Song and Air Call
32 Feral Pigeon Song and Call
33 Medley: Woodpigeon song, Collared Dove song (after 20 seconds), Feral Pigeon song, Woodpigeon (again, as the track draws to a close).
34 Chaffinch Song
35 Chaffinch Calls
36 Goldfinch Call and Song
37 Greenfinch Call and Song
38 Medley: Chaffinch song, Greenfinch song (at 25 seconds), Goldfinch calls (at 38 seconds), Chaffinch (final verse at the end).
39 House Sparrow Call and Song
40 Starling Song
41 House Martin Call and Song
42 Swift Call
43 Pied Wagtail Calls
44 Grey Wagtail Calls
45 Medley: Pied Wagtail calls, Starling calls, House Martin call (at 28 seconds), Swift calls, House Martin calls (alongside the Swifts), Grey Wagtail calls, House Sparrow calls (at 43 seconds). Listen for the Grey Wagtail calls in the background.
46 Carrion Crow
47 Jackdaw
48 Magpie
49 Medley: Robin song, Collared Dove song, Greenfinch song, Magpie calls, Wren song, Carrion Crow calls, Woodpigeon song, House

Sparrow calls, Jackdaw calls, Great Tit song, Blackbird song, Blue Tit song, Dunnock song, Goldcrest song.

50 Song Thrush Song

51 Mistle Thrush Song

52 Mistle Thrush Call

53 **Medley:** Mistle Thrush song, Song Thrush song, Blackbird song (after 32 seconds).

54 Chiffchaff Song

55 Willow Warbler Song

56 Blackcap Song

57 Chiffchaff, Willow Warbler and Blackcap Calls

58 Garden Warbler Song

59 Nightingale Song

60 **Medley:** Blackcap song, Willow Warbler song, Nightingale song, Blackcap song, Willow Warbler song, Nightingale song (twice), Garden Warbler song. Listen for the Chiffchaff singing in the background throughout much of the track.

61 Great Spotted Woodpecker Drum and Call

62 Nuthatch Calls

63 Green Woodpecker Yaffle

64 Jay Call

65 Buzzard Call

66 Marsh Tit Calls

67 Tawny Owl Hoot and Call

68 Long-tailed Tit Calls

69 Treecreeper Song and Call

70 Bullfinch Call

71 Stock Dove Song

72 Spotted Flycatcher Call

73 **Medley:** Marsh Tit calls, Willow Warbler song, Great Spotted Woodpecker drum, Treecreeper calls, Stock Dove song (at 16 seconds), Jay call (at 26 seconds), Bullfinch calls, Long-tailed Tit calls, Nuthatch song, Chiffchaff song.

74 Wood Warbler Song

75 Pied Flycatcher Song

76 Redstart Song and Call

77 Tree Pipit Song

78 **Medley:** Alternating Pied Flycatcher and Redstart song, Wood Warbler song, Redstart song, Tree Pipit song, Wood Warbler song, Tree Pipit song (again).

79 Rook

80 Skylark

81 Linnet

82 Yellowhammer

83 Corn Bunting

84 Whitethroat Song

85 Lesser Whitethroat Song

86 Swallow Song

87 Cuckoo Song

88 Meadow Pipit Song and Call

89 Pheasant Advertising Call

90 Red-legged Partridge Advertising Call

91 Grey Partridge Advertising Call

92 **Medley:** Yellowhammer song, Whitethroat song (in between), Corn Bunting song, Cuckoo song (after 17 seconds), Red-legged Partridge calls (in the distance), Woodpigeon song, Linnet song (at 25 seconds), Skylark song (at 28 seconds), Pheasant's call (at 30 seconds), Rooks (in the background), Lesser Whitethroat (at 36 seconds), Whitethorat (again), a Swallow call (at 45 seconds). A Blackbird can be heard in the background.

93 Reed Bunting Songs

94 Reed Warbler Song

95 Sedge Warbler Song

96 Cetti's Warbler Song

97 Grasshopper Warbler Song

98 **Medley:** Reed Warbler song, Carrion Crow call (in the background), Cetti's Warbler song (at 16 and 24 seconds), Reed Bunting (at 19 and 30 seconds), Sedge Warbler song (at 20 seconds), Grasshopper Warbler song (at 33 seconds).

99 Over to You

Index

Detailed descriptive entries for birds are indexed in **bold**. Scientific names are indexed with main page number only.